SCIENCE EDUCATION AND CITIZENSHIP

HISTORICAL STUDIES IN EDUCATION

Edited by William J. Reese and John L. Rury

William J. Reese, Carl F. Kaestle WARF Professor of Educational Policy Studies and History, the University of Wisconsin-Madison.

John L. Rury, Professor of Education and (by courtesy) History, the University of Kansas.

This series features new scholarship on the historical development of education, defined broadly, in the United States and elsewhere. Interdisciplinary in orientation and comprehensive in scope, it spans methodological boundaries and interpretive traditions. Imaginative and thoughtful history can contribute to the global conversation about educational change. Inspired history lends itself to continued hope for reform, and to realizing the potential for progress in all educational experiences.

Published by Palgrave Macmillan:

Democracy and Schooling in California: The Legacy of Helen Heffernan and Corinne Seeds,
by Kathleen Weiler

The Global University: Past, Present, and Future Perspectives,
edited by Adam R. Nelson and Ian P. Wei

Catholic Teaching Brothers: Their, Life in the English-Speaking World, 1891–1965
by Tom O'Donoghue

Science Education and Citizenship: Fairs, Clubs, and Talent Searches for American Youth, 1918–1958
by Sevan G. Terzian

SCIENCE EDUCATION AND CITIZENSHIP

Fairs, Clubs, and Talent Searches for American Youth, 1918–1958

Sevan G. Terzian

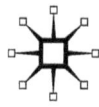

SCIENCE EDUCATION AND CITIZENSHIP
Copyright © Sevan G. Terzian, 2013.

All rights reserved.

First published in 2013 by
PALGRAVE MACMILLAN®
in the United States—a division of St. Martin's Press LLC,
175 Fifth Avenue, New York, NY 10010.

Where this book is distributed in the UK, Europe and the rest of the world, this is by Palgrave Macmillan, a division of Macmillan Publishers Limited, registered in England, company number 785998, of Houndmills, Basingstoke, Hampshire RG21 6XS.

Palgrave Macmillan is the global academic imprint of the above companies and has companies and representatives throughout the world.

Palgrave® and Macmillan® are registered trademarks in the United States, the United Kingdom, Europe and other countries.

ISBN: 978–1–137–03186–0

Portions of chapter 3 appeared in a different form in "The 1939–1940 New York World's Fair and the Transformation of the American Science Extracurriculum," *Science Education* 93 (September 2009): 892–914. Permission to include text from that article was granted as a courtesy of John Wiley & Sons, Inc. Publishers.

Portions of chapter 5 appeared in a different form in "'Adventures in Science': Casting Scientifically Talented Youth as National Resources on American Radio, 1942–1958," *Paedagogica Historica* 44 (June 2008): 309–325. Permission to include text from that article was granted as a courtesy of Taylor & Francis publishers.

Permission to include text and illustrations from the "American Institute of the City of New York for the Encouragement of Science and Invention, MS 17" collection was granted as a courtesy of the New-York Historical Society.

Permission to include illustrations from the "Science Service Records, RU 7091" collection was granted as a courtesy of the Smithsonian Institution Archives.

Permission to include illustrations of Science Talent Search finalists was granted as a courtesy of Society for Science and the Public.

Permission to include text from "The Papers of Harlow Shapley," (HUA HUG 4773.10) was granted as a courtesy of the Harvard University Archives.

Library of Congress Cataloging-in-Publication Data is available from the Library of Congress.

A catalogue record of the book is available from the British Library.

Design by Newgen Imaging Systems (P) Ltd., Chennai, India.

First edition: January 2013

10 9 8 7 6 5 4 3 2 1

Transferred to Digital Printing in 2013

For Lee-Ann

Contents

List of Figures and Tables	ix
Foreword	xi
Acknowledgments	xiii
Introduction	1
1 Origins of Science Clubs and Fairs	9
2 Building a Network	33
3 Showcasing Young Scientists at the New York World's Fair	57
4 Enlisting Science Education for National Strength	81
5 Sustaining Mobilization in an Atomic Age	103
Conclusion	139
Notes	147
Selected Bibliography	211
Index	227

Figures and Tables

Figures

2.1	The American Institute's Science Congress program cover	41
3.1	Girls perform chemical experiments in the Westinghouse building	67
3.2	Poster promoting the American Institute Science and Engineering Clubs' exhibits	69
3.3	Shirley Gesser prepares microscopic slides	77
5.1	Science Talent Search winners accompany Senator Harley Kilgore	107
5.2	Science Talent Search winners join Science Service staff writer Frank Thone	117
5.3	The 40 Science Talent Search winners visit US President Harry S. Truman	119
5.4	J. Robert Oppenheimer, A. C. Monteith, Watson Davis, and Harlow Shapley congratulate two Science Talent Search winners	120
5.5	Science Clubs of America brochure	128

Tables

2.1	Children's Science Fair projects by subject	46
2.2	Seventh, eighth, and ninth grade boys' and girls' exhibits	47
2.3	Tenth, eleventh, and twelfth grade boys' and girls' exhibits	47
5.1	Science Talent Search winners by gender	133

Foreword

The extra curriculum has long been a neglected topic in the history of education, but in this thoroughly researched volume Sevan Terzian demonstrates its potential importance as a vital dimension of school life and education policy. Science fairs and clubs have become a familiar aspect of school life in the United States, but they have a history that touches upon such larger questions as the role of science in public life and its contributions to national security. Terzian carefully delineates these questions, contrasting arguments about the role of science education in promoting a more democratic society with more utilitarian concerns about the country's readiness to compete in a technologically complex world. These competing visions animated discussions for decades during the mid-twentieth century of how best to organize the teaching and learning of the sciences. In the end it, however, was the question of global competitiveness that proved most prominent, and its influence was evident in the development of science clubs, science fairs, and related activities in schools across the nation.

Extra curricular activities are often thought to be the province of student organizations and teen enterprise, but Terzian reveals that they can also become a focal point of adult interest as well. This was certainly true in the case of science fairs and clubs, which were planned and promoted by influential educators and scientists, as well as professional associations and advocacy groups. Terzian's scholarship highlights the role of New York City as an early focal point of these efforts, offering a potent model for the rest of the country. With visionary educators such as Morris Meister and S. Aleta McEvoy leading the way, and organizations such as the American Institute of the City of New York providing an infrastructure and other sources of support, a national movement to use the extra curriculum to advance the cause of science education gradually gained momentum. Key faculty members in university schools of education also provided leadership, and eventually critical support came from commercial sponsors, particularly the Westinghouse Corporation. This was a rare and early case of using an educational activity as a corporate branding strategy, and one that appears to have been quite successful. Science was far too important an issue to be left to the kids, and supporting clubs and fairs provided a publicly laudable means of fostering an image of expertise and innovation. By the mid-twentieth century science had become a big business, and Terzian documents a critical aspect of its impact on the schools.

Given these developments, along with the advent of Cold War tensions, it is little wonder that science clubs and fairs became associated with national defense and economic competitiveness by the 1950s. Judging from information that Terzian provides, they also were predominantly white and male with respect to the students who participated. This did not occur by design, but it certainly reflected accepted stereotypes about scientists depicted in the media and popular culture at the time. Despite rhetoric about the democratic quality of the clubs and fairs, there were only sporadic efforts to involve girls and students from ethnic and racial minority groups. If science was indeed a vital aspect of the country's economic future, it was not likely to offer a pathway to success for these students. As Terzian notes, this was yet another shortcoming of the movement. Despite its eventual growth and influence, it was a dimension of educational change that had little bearing on questions of equity and enhancing opportunity for groups historically excluded from the scientific enterprise.

In each of these respects and more, Sevan Terzian has provided us a thoughtful glimpse into the past, carefully documenting the development of this long-neglected facet of American education in the twentieth century. We are pleased to include this volume in the series, and hope that it will move other scholars to examine additional facets of the extra curriculum that has shaped the informal education of Americans. As Terzian has established, such activities were hardly a trivial aspect of the school experience, and occasionally became the object of debate and conflict among adults. Such history can only help to deepen our appreciation of how these processes continue to operate in schools and other educational settings today. Helping to pose such questions about the present and future makes this a most useful history indeed.

WILLIAM J. REESE
AND
JOHN L. RURY

Acknowledgments

The generous assistance and support of many people have helped make this book possible. My editor at Palgrave Macmillan, Burke Gerstenschlager, conveyed enthusiasm and helped move this project forward. Kaylan Connally at Palgrave has been especially responsive in helping me navigate the path to publication. I am also delighted that this book joins the Historical Studies in Education series coedited by William Reese and John Rury, whose scholarship I have long admired.

I have had the privilege of making the University of Florida my professional home for the past 12 years. I am grateful to many colleagues for their encouragement and thoughtful criticism, especially Elizabeth Bondy, Kent Crippen, Jean Crockett, Tom Dana, Harry Daniels, Jack Davis, Kara Dawson, Cynthia Griffin, William Link, James McLeskey, Stephen Pape, Richard Renner, Dorene Ross, Troy Sadler, Paul Sindelar, Joseph Spillane, Dave Tegeder, and Robert Zieger. I have also benefited enormously from the keen insights of many current and former graduate students affiliated with social foundations of education at the University of Florida, including Donald Boyd, Christopher Brkich, Colleen Butcher, Jessica Clawson, Kathryn Comerford, Robert Dahlgren, Chanel Gaiter, Regan Garner, Erika Gubrium, Andrew Grunzke, Sheryl Howie, Emma Humphries, Amy Martinelli, Steve Masyada, Leigh Ann Osborne, Patrick Ryan, Leigh Shapiro, Katie Tricarico, and Lauren Tripp. Donald Boyd, Jessica Clawson, and Patrick Ryan also were exceptionally capable research assistants.

Financial support for travel to the sources and time away from teaching proved to be especially beneficial. A research grant from the Spencer Foundation (Grant # 200900000) allowed timely support for analysis and writing. Two College Research Incentive Fund awards from the College of Education at the University of Florida helped cover costs related to archival research. A sabbatical leave from the University of Florida during the spring 2008 semester was critical for conceptualizing and writing the first two chapters.

Like all historians, I have leaned heavily on the expert guidance of librarians and archivists. Ted O'Reilly helped make two trips to the New-York Historical Society exceptionally productive. Cassandra Nespor at the Senator John Heinz History Center in Pittsburgh helped me navigate the collections pertaining to Westinghouse. Amy Crumpton at the American Association for the Advancement of Science directed me to State Academy and Junior Academy of Science materials. Librarians at the Harvard University Archives granted me access to the Harlow Shapley papers. Special thanks also go to Rick Bates, Norma Rosado Blake, Reynolds Clark, Jim Davis, Eleanor Gillers, Cheryl Kubelick, Art Louderback, and Tammy Peters. Most of all, I am deeply indebted to Ellen Alers at the Smithsonian Institution Archives, whose cheerful, insightful, and patient direction allowed me to take full advantage of the Science Service records on many occasions.

I have also relied on the generosity of friends and family. Tamar Terzian and Joe Masih hosted me while on research trips to Washington, D.C., as did Monty Cox and Robyn Holtzman. Ken and Rosalie Kizirian graciously welcomed me to their home in Boston. In New York City, Eric Slosberg and Sarah Levinson helped make my visits especially productive and enjoyable. I imposed on the hospitality of Leo Peña on multiple occasions while sharing my research at conferences in Chicago. I am grateful to all of them for their interest and support.

I have found the History of Education Society and the History & Historiography Division of the American Educational Research Association to be exceptionally stimulating venues for sharing (and improving) my research. Special thanks go to Randall Curren, Linda Eisenmann, Barry Franklin, David Gamson, Jeffrey Mirel, Amy Slaton, and Kim Tolley for their careful and constructive criticism of earlier versions of this work. Linda Laffey and Caroline Kelsey helped me select images to include in this volume. Bruce Lewenstein's interest and enthusiasm about this project have encouraged me from the outset. Marcel LaFollette's unparalleled generosity, keen insight, and collegiality have been nothing short of inspirational. Karen Graves, Lee-Ann Laffey, William Link, and John Rudolph each read an earlier version of the entire manuscript, and I am indebted to each of them for their detailed, constructive, and critical assessments. I continue to aspire to their high standards of historical scholarship.

I was raised in a family in which science promised both wonderful discoveries and ways of thinking that could improve human life.

My father, Yervant Terzian, and mother, Araxy Bablanian, have long championed the rigor and beauty of scientific thought. Although my academic studies eventually gravitated to the humanities, my appreciation of science and science education remained and eventually led me to this historical investigation.

Most of all, my wife, Lee-Ann Laffey, and our children, Talar and Christopher, have helped make this project especially fulfilling. Lee-Ann is my closest companion, wisest adviser, and constant source of encouragement and inspiration. I dedicate this book to her.

SEVAN G. TERZIAN
Gainesville, Florida
May 2012

Introduction

On November 4, 1931, Hugo Newman, Principal of New York Teacher Training College, spoke on NBC radio about the importance of science fairs in American society. "Our growing generation will be better prepared in the art of living their lives happily and successfully," Newman declared. Science fairs offered students unprecedented opportunities "to widen their horizon, to increase their knowledge of the living world about them, to take an active part in the solution of their problems, and to acquire and foster that most desirable and productive quality—'the scientific habit of mind.'"[1] In April 1951, W. Stuart Symington, Chairman of the National Security Resources Board, endorsed science fairs for a different reason: "Scientific and technical know-how have made this Nation a leader among nations, and will keep it so." He urged educators and newspapers to continue supporting science fairs to remedy a "critical shortage of scientific personnel." "With this sort of watchful leadership," Symington concluded, "America will never be caught technically unprepared."[2]

Spanning years of economic depression, world war, and an atomic age, these two statements convey markedly distinct civic values. For Newman, a veteran science educator and pedagogical innovator, science fairs and science club activities exposed students to new ideas about the natural world. Participants became acquainted with their local surroundings, developed reasoning powers, and learned how to address and resolve problems of mutual interest. In short, science fairs cultivated productive dispositions and fruitful ways of living. For Symington, the United States' political power in an atomic age derived from and depended upon science. Student achievement through science fairs would ensure that new generations of technological experts would secure the nation's privileged status. Professional educators could be trusted to groom talented specialists who would fortify the United States militarily and economically. For Newman, science fairs and clubs heightened students' social awareness and elicited their involvement in community affairs. Symington, meanwhile, believed that science fairs oriented youth to national imperatives.

These contrasting viewpoints prompt a series of questions. Where and why did science fairs and clubs originate? What roles did science educators and professional scientists play in fostering this youth movement? In the early 1940s, an annual science talent search began to identify and reward a new generation of technological experts. Who engineered that campaign, and why did it assume political urgency on a national scale? Furthermore, in what ways did larger social and political concerns inform the civic justifications for these educational programs: From the time of their inception after World War I to the national panic over *Sputnik* during the Cold War? Many Americans are familiar with science fairs, clubs, and talent searches, as they have become fixtures in the annual rituals of schools. Yet relatively little is known about their origins, purposes, and proliferation.[3] That is the subject of this book. In focusing on the civic justifications for these extracurricular programs, this historical investigation also considers the uneasy juxtaposition of democratic and meritocratic ideals in American education.

Initially, groups of science teachers, scientists, and popularizers of science promoted clubs and fairs to nurture participatory democracy in local communities. Under the direction of professional educators in decades marked by fears of juvenile delinquency and economic crisis, these programs aimed to develop constructive social habits, powers of reasoning, active inquiry, and an awareness of the natural environment. Such skills and dispositions, proponents argued, would create citizens who could evaluate and appreciate the importance of science in public affairs. A democracy also required citizens who could think like scientists in evaluating and devising rational solutions to societal problems. Science clubs typically emerged in urban secondary schools, where relatively high enrollments and the availability of science courses often yielded a critical mass of interested students. By the late 1920s, children began to display their projects and compete for prizes in public forums. These were the first science fairs, the most elaborate of which flourished in New York City. A host of communities in Northeastern and Midwestern states subsequently launched their own. Despite educators' lofty rhetoric about empowering future generations of citizens, however, only a small fraction of American schoolchildren had opportunities to join these voluntary organizations. Even as the movement spread, and as high school enrollments ballooned in the 1920s and 1930s, many students lacked access to science courses that could spark their interest in clubs and fairs.

By the early 1940s, the nation's mobilization for World War II had transformed these educational programs. Fairs, clubs, and a newly created Science Talent Search addressed problems of national defense. Science educators, political leaders, and industrial patrons urged that the United States' military and economic strength depended upon the systematic identification and training of talented youth who would ultimately develop weapons and other commodities to secure Allied victory. Searches for the most promising minds and practical innovators would exemplify meritocratic ideals. The democratic rationale became secondary, as primarily the most intellectually gifted students were sought. In the ensuing atomic age and Cold War, the quest to locate and reward future scientific elites persisted through nationally coordinated programs including the talent search, Science Clubs of America, and the National Science Fair. The mobilization of science education for national defense thus began during World War II and well preceded Americans' responses to the Soviet Union's launching of *Sputnik* in 1957. Patterns of student participation and achievement in these programs, however, reflected pervasive inequalities in American secondary schooling. Despite their stated civic purposes, science education in these instances was neither democratic nor meritocratic.

Historians of education have long questioned the extent to which American public schools embodied and promoted democratic values. From their origins in the nineteenth century, common schools assumed a civic mission, but rarely encouraged universal political participation. More typically, they aimed to cultivate morally virtuous citizens who upheld a stable republic. By the early twentieth century, rising industrialization and immigration, coupled with booming enrollments and an increasingly diverse student body, prompted public schools to introduce a vocational dimension that sometimes superseded this original civic purpose. Some concurrent public reforms in this era, meanwhile, expanded the rights of citizens, while others were designed to limit direct political participation and cement social stability.[4]

In the midst of these developments, the architects of science clubs and fairs promoted deliberative democracy in ways that reflected American philosopher John Dewey's educational ideals.[5] For Dewey, democratic society transcended government; it was "a mode of associated living of conjoint communicated experience."[6] As a result, schools should help students become aware of their common interests, interdependence, and limits to their individual freedom. To be inclusive, formal education assumed responsibility for nurturing all students' inclinations to participate actively in their learning. Schools also taught people how to build and strengthen communities by reconciling their personal preferences with the needs of larger society—"an organic union of individuals."[7] This mediation required "habits of mind which secure social changes without introducing disorder."[8] As an epistemological pragmatist, moreover, Dewey believed that knowledge and society were constantly changing. Social problems should therefore be approached "experimentally." For democratic deliberation to proceed earnestly and without destructive conflict, people needed to eschew impulsive behaviors by anticipating the consequences of their actions. In that sense, citizens should emulate a scientific investigator who shed personal biases, weighed alternative ideas, and accepted the definitive authority of nature.[9]

At the same time, much American educational thought has embraced meritocratic ideals in explaining social differences and economic inequalities in a democracy. Because all are presumed to enjoy the same opportunities to achieve, the allocation of unequal rewards across the population is widely accepted. A meritocratic society ignores ascriptive characteristics such as a member's race, ethnicity, socioeconomic class, or sex—none of which is presumed to bear on academic or job performance. Schools, then, must sort students solely according to their educational achievements and attainments. In practice, however, ascriptive characteristics have often shaped student behavior and achievement, while schools struggle to measure talent and achievement consistently and in ways that reliably predict future accomplishments.[10]

In the early twentieth century, singular, innate, and hereditary notions of intelligence appealed to many American scientists, social scientists, and educators, because they promised efficient and objective methods for classifying students in an increasingly mechanized society. Nonetheless, educational institutions often reproduced societal inequalities—albeit in a meritocratic guise.[11] Narrowly defined notions of merit, for instance, systematically excluded many racial minorities from educational advancement and career opportunities in the engineering sciences. Particularly in technical fields that were seemingly shielded from social or political biases, American schools frequently adopted meritocratic mechanisms and aims in ways that perpetuated racial and other societal inequalities.[12] According to historian John L. Rudolph, moreover, the emergence of a Cold War political ideology, "laissez-faire individualism," and a robust economy in the late 1940s and 1950s fortified meritocratic values in American education.[13] In the realm of science education, the nation's most talented and accomplished students had a duty to aspire to scientific careers. Their subsequent research and applications, many believed, would determine the United States' military security and economic prosperity. With these heightened stakes, American science education assumed meritocratic mechanisms and purposes. Democratic aims did not convey the same degree of urgency.

This investigation of the civic dimensions of science fairs, clubs, and talent searches looks beyond the curriculum in featuring a significant, yet largely neglected, aspect of American science education.[14] From this vantage point, World War II emerges as a transformative event. Most historians of education have devoted insufficient attention to the significance of World War II, and they have largely underestimated its enduring impact on American schools. This history of science fairs, clubs, and talent searches reveals that the mobilization of American schools for national defense that began during World War II persisted well into the postwar era.[15] The nation's political imperatives oriented this aspect of American education in the 1940s and 1950s.[16] Patterns of student participation and achievement in science, meanwhile, indicate gender, racial, and regional inequalities.[17]

The main narrative encompasses five chapters. The first begins by delineating the initiatives of Morris Meister, an energetic and widely influential science educator from New York City. It highlights Meister's articulation of the civic and social benefits of the project method and after-school science clubs in terms congruent with Dewey's pedagogical and democratic aims. Science clubs emerging in various communities during the 1920s often emulated these purposes. Simultaneously, some state affiliates of the American Association for the Advancement of Science (AAAS) created Junior Academies of Science, in which professional scientists and high school science teachers included students in annual meetings and acquainted them with the rigors of the field. In New York City, meanwhile, a pioneering annual event for youth was underway: the Children's Science Fair. Sponsored by a century-old organization for the promotion of science and industry—the American Institute of the City of New York—these science fairs began in 1928 as a way to facilitate students' firsthand inquiry of the natural world. All of these initiatives prized the civic virtues of science education and eschewed explicitly vocational aims. They sought to develop students' powers of reasoning through active inquiry—skills and dispositions for their roles as democratic citizens.

Science clubs and fairs proliferated in the 1930s. As chapter two illustrates, these activities in New York City grew substantially, and they arose in other parts of the nation despite the economic constraints of the Great Depression. The American Institute created a series of programs to acquaint students from overcrowded schools with investigative methods of scientific inquiry. These included weekend demonstration lectures and workshop courses at the American Museum of Natural History, New York University, and other local venues. An annual Science Congress, modeled after AAAS meetings, featured dozens of student experiments and lectures from prominent scientists. These initiatives proved to be costly, but immensely popular. In the process, a citywide network of students, science teachers, museums, and institutions of higher education had emerged. American Institute officials claimed that they groomed productive social habits and empowered youth in an era of economic hardship and social instability. Beyond New York City, Junior Academies of Science continued to sponsor high school science clubs and annual conventions. Although they lacked some of the elaborate features of club activities in New York City, these programs often followed the American Institute's precedents and embodied a similar civic mission. By the late 1930s, some began to envision a nationally coordinated youth movement in science.

Chapter three locates a key moment of transition. Aspiring to expand its science education programs nationally, the American Institute secured the financial support of a leading industrial corporation: the Westinghouse Electric and Manufacturing Company. Organizers of the science club activities in New York City also arranged to showcase the projects and experiments of distinguished local students in Westinghouse's building at the New York World's Fair from 1939 to 1940. This collaboration ultimately proved problematic, however, as conflicts erupted about the social and political values of science. Science educators touted the merits of technical expertise and the processes of active inquiry, while the corporate sponsors sought to orient World's Fair visitors to the entertaining dimensions and consumer products of science. In the fair's aftermath, Westinghouse terminated its financial support of the American Institute, which severely curtailed science education activities in New York City. Instead, Westinghouse invested in a longstanding agency for science popularization in the nation's capital: Science Service, Inc.[18] As the United States began to mobilize for war, moreover, science clubs and fairs oriented their activities to national defense. The quest to cultivate critically thinking citizens in a democracy had diminished.

World War II introduced new meritocratic justifications for science education in the United States. In the summer of 1941, Science Service's director, Watson Davis, and Westinghouse's executives created a national competition for high school seniors: an annual Science Talent Search. Aiming to conserve natural resources for Allied victory, these programs sought to identify and enlist select youth for national strength. Chapter four discusses these initiatives and the political messages students in science clubs and talent search winners received from professional scientists, politicians, and military leaders about their importance in helping the United States win the war. By 1945, the talent search's meritocratic aims influenced the federal government's recommendations for a postwar national science program in the US Office of Scientific Research and Development's report, *Science—The Endless Frontier*. With this sort of attention, and rapidly increasing student membership, Science Service's leaders would continue to address the nation's military and economic needs after the war ended.

These science education programs retained their meritocratic purposes and nationalistic orientation in the postwar era as a reflection of the United States' material prosperity and sustained military mobilization. Chapter five highlights the power of the Cold War political ideology in prompting an unprecedented rise in the number of science clubs across the nation. By 1950, a new annual competition for high school students—the National Science Fair—urged participants to apply their talents to the nation's military and technological prominence. Junior Academies of Science, which furthered their expansion from the 1930s, adopted comparable political aims. The United States' rivalry with the Soviet Union also encouraged Science Talent Search participants to aspire to roles of expert leadership in an atomic age. Despite the quest to locate and reward the most promising youth indiscriminately, however, girls, racial minorities, and rural youth were perennially underrepresented. Students' unequal opportunities to compete and succeed compromised the talent search's meritocratic mission and reflected existing inequalities in American secondary education. As the concluding chapter demonstrates, the historical pursuit of talented youth to fortify the nation militarily and economically anticipated more recent calls for reforming science education in the United States. Ultimately, the transformation of American science education's civic purposes in the mid-twentieth century was emblematic of the United States' emergence as the world's leading political power during and following World War II.

CHAPTER 1

Origins of Science Clubs and Fairs

In the midst of American military involvement in World War I, a young participant at a conference on science education at Teachers College in New York City opened his address with a bold declaration: "The war is the most vital factor in the world today. America is the most vital factor in the war. Education is the most vital permanent factor in America. Science, considered in the large, can and must become the most vital factor in Education." The speaker then pointed to science education as the most promising means for societal progress: "We, to whom has been entrusted this dominant note of modern life, are now confronted with the golden opportunity for change which comes with every cataclysm in life."[1] This was the conviction of Morris Meister, a doctoral student at Columbia University and science teacher in the New York City public schools. His teaching and research over the ensuing decade pioneered efforts to engage students in hands-on experiments in the science classroom, and increasingly, in after-school science clubs. Believing that schools must "utilize the social nature of the boy and make each and every subject a part of the real life of the school," Meister developed and promoted an elaborate rationale and plan for science clubs. Science had rapidly yielded scores of technological innovations that shaped modern living. As a result, science clubs would empower future generations by educating "a citizenship of men and women really appreciative and intelligent in judging the affairs of life."[2] His application of John Dewey's educational philosophy to science activities for youth would influence thousands of teachers in New York City and across the United States in the 1920s and early 1930s.[3]

Meister taught science at the new and innovative Speyer Junior High School from 1916 to 1918 and then at Horace Mann Elementary School until 1921, when he earned his doctoral degree. In both his studies and teaching, Meister experimented with new pedagogies to make science more meaningful and influential in the lives of youth.[4] Like many educators of his generation, Meister believed that public schools should extend their reach to combat the seemingly pernicious effects of increased leisure time: "What our pupils do during every hour of the twenty-four in the day—and of every day in the year—is...a legitimate consideration for the school and teacher."[5] His doctoral dissertation, "The Educational Value of Certain After-School Materials and Activities in Science," analyzed the rising popularity of toys and the companies that produced and marketed them to American youth. Concerned by "the taint of commercialism [that] endangers the whole future of the boy science movement in this country...[and]...puts the child at the mercy of the sale manager of the company," Meister endeavored to understand why youth activities affiliated with toy manufacturers were popular. Subscriptions to company magazines or booklets, he believed, fostered a sort of group ethos or "spirit" that presented an "outlet for the 'gang' instinct or tendency among boys of a certain age."[6]

Indeed, widespread concerns about American youth were intensifying in the decade following World War I. The advent of a modern society, particularly in urban settings, had introduced movie theaters, telephones, and automobiles. Many adults worried that increased leisure time, coupled with the allure of commercial entertainment, prompted the rise of peer groups and dating, which, in turn, weakened moral traditions and diminished parental influence. Seeking immediate and material gratification in an era scarred by the horrors of the world war, the behavior and attitudes of younger generations led some critics to lament that American society had become conducive to juvenile delinquency. Some speculated that child rearing was no longer instinctive, and they implicated parents for raising thieving boys and sexually promiscuous girls. In response, many professional educators encouraged longer school days and terms for youth under their supervision. The celebrated *Cardinal Principles of Secondary Education* report of 1918 reflected this conviction in establishing "worthy use of leisure," "worthy home-membership," and "citizenship" among its seven central purposes.[7]

In trying to widen the influence of schools in the lives of American youth, Meister sought to understand the appeal of commercial organizations to younger generations. Specifically, he felt that toy companies' competitions for scholarships proved popular with boys who wanted to measure their worth against one another. "Boy Universities," like the International Society of Meccano Engineers, the Gilbert Institute of Engineering, or the Boys' Chemcraft Chemist

Club, fostered feelings of belonging. Letter bureaus of toy company magazines similarly held enormous appeal. In their ability to capture the developmental interests of early adolescent boys, Meister considered these commercial organizations to be a sort of model for creating after-school science clubs: "To bring the boys of the whole country together in this common pursuit, with this common interest and in cooperative effort is an ideal which can take the shape of a real boy movement in the field of science."[8] However, because the effects of affiliating with these for-profit institutions were unclear, and as the lack of personal contact from mail correspondence was limiting, activities under teacher supervision posed a better alternative. Rather than allow toy companies to influence youth for their own gains with their "advertising propaganda, so genuine and keen an interest should legitimately fall to the teacher for development."[9] Like many social reformers of his generation, Meister aimed to bolster and extend the authority of public schools.

Junior high schools appeared especially suited to the task. As nascent additions to the public school system, particularly in urban districts, they were open to innovations such as ability grouping, interdisciplinary curricula, and new pedagogies such as the project method. From a developmental standpoint, according to Meister, junior high schools could accommodate the "gang tendencies" of early adolescent boys, as they also tended to enjoy tinkering with machines. Specifically, after-school science clubs, featuring projects and first-hand experiments, could accommodate students' inclinations for shop and laboratory work.[10]

More generally, Meister envisioned an expansive program of general science for all elementary and junior high school students. Regardless of whether a student aspired to become a scientist, science education should "enable our pupils to appreciate the methods of science and to use this method and the thinking procedure of science in their every-day lives."[11] Like many educators of his generation, Meister touted the civic and social benefits of understanding how "scientific laws and principles" applied to business and governmental affairs. Scientific knowledge also held "avocational value," or improved aesthetic awareness of one's natural environment. Training future professional scientists assumed secondary importance.[12] Meister's priorities reflected some of the emerging justifications for science education in the early twentieth century. For example, in 1920 the Commission for the Reorganization of Secondary Education's science committee emphasized the importance of teaching students the value of good health, responsible citizenship, and productive uses of leisure time. The rise of general science courses in the interwar decades, especially in the eighth and ninth grades, would attempt to ensure that a wider segment of youth would gain at least some exposure to the field and its societal applications.[13]

In secondary education, Meister believed that the project method could capture and sustain students' curiosity about science while furthering the school's social functions. Historians have demonstrated that the project method popularized by William Heard Kilpatrick did not always remain faithful to the philosophical tenets of John Dewey's educational progressivism. According to Kilpatrick's conception, the project method catered to students' particular interests by allowing them the opportunity to study the academic subjects of their choosing. It tended to emphasize the practical applications of student activity to the point where a teacher's primary responsibility was reduced to directing that work to purposeful ends. A teacher's intellectual mastery of subject matter diminished in importance. Dewey, by contrast, cast the teacher as a powerful moral and academic guide. Furthermore, there was a larger civic end to active learning, interdisciplinary activities, and individually meaningful lessons: cultivating both rational thought and empathy for healthy citizenship and societal progress. Despite his zeal to promote Dewey's philosophy, Kilpatrick's brand of educational progressivism tended to lose sight of that overarching goal.[14]

In the realm of science education, Dewey found a more faithful advocate in Meister, whose conception of the project method reconciled both pedagogy and civic purpose.[15] Meister had implemented the project method in the science classes he taught at the Speyer School, because it accommodated students' interests. But he also stressed its social and ethical dimensions. Unlike many science educators who, as historian John L. Rudolph has shown, unduly simplified Dewey's stages of thought to a "formulaic" and "numbingly algorithmic" arrangement called "the scientific method,"[16] Meister encouraged students to tinker with any number of everyday machines and chemicals. He echoed Dewey's mantra in declaring that the project method was "closely allied with the spirit of democracy in education." Meister promoted scientific understanding among all students to cultivate new generations of citizens capable of evaluating and resolving societal problems. Although he would later found one of the most selective and competitive magnet schools for science, Meister's quest to promote widespread scientific literacy remained prominent throughout his career.[17] "The world won't solve its problems," he later reflected, "until the masses understand the world they live in." This meant that the vast majority must comprehend scientific principles and their societal consequences: "Until ninety percent of the people know the workings of the human body, we will not have a healthy people. Until ninety percent of the people know the composition of the crust of the earth, we will not be able to make full use of the materials in it." "The world can be made a better place," Meister concluded, "through scientific education."[18]

Teachers had the responsibility of establishing parameters and directing students' projects. In this, Meister echoed Dewey's recognition of an apparent paradox in democratic life: "The highest kind of freedom of the individual is built upon a system of controls and restrictions—so in a project there must be a definite place for guidance and control."[19] He articulated Dewey's criteria for evaluating the worth and suitability of student projects. Science teachers therefore had the responsibility of determining whether a student's interest in a proposed project was genuine, purposeful, and feasible: "It is a mistake to think that in a project all that the teacher does is to get out of the way."[20] Frequent consultations would allow teachers to propel, navigate, and monitor students' science projects. Teachers also needed to furnish classroom resources including an extensive library of science books, magazines, newspapers, and pamphlets; a file of productive science projects, experiments, and related questions; and a school shop for laboratory experiments. To connect this individualized activity to larger social ends, Meister established detailed criteria for each student to prepare an interactive lecture or report to the class. The teacher, having closely guided the project to fruition, then allowed the student to lead this "socialized recitation."[21] By balancing teacher authority with student initiative, Meister believed, these sorts of "play activities" or laboratory projects leading to class demonstrations facilitated students' direct contact with natural phenomena. Such purposeful experiences empowered students by reducing their dependence on mechanical or electrical experts and, more generally, made "the average individual a less gullible, more inquiring, and better-reasoning citizen."[22]

Science clubs complemented Meister's implementation of the project method in the curriculum. He viewed after-school organizations as a ripe arena for teachers to influence students' leisure activities. With the proper balance of control and freedom, teachers could supervise science clubs "without losing that free, vital, purposeful urge to thought and to action that is so common to things our pupils do outside the classroom and often so sadly lacking in the things they do for us during school hours."[23] Meister organized a science club at the Speyer School to stimulate greater student enthusiasm: "It seemed entirely contrary to the supposed interests of boys that there should be a flourishing Latin Club in the school, among other things, and no interest in science."[24] Taking inspiration from his work in New York City's settlement houses and recreation centers, moreover, Meister viewed school clubs as a "safety valve" for impressionable boys in the "gang age." It was especially important to him that science clubs contribute to the overall welfare and spirit of the school community.[25]

At the Speyer School, for instance, Meister attempted to align the science club with the school's larger mission by outlining a "creed" of character traits and skills that each member should possess. Only 20 students qualified as "grade A" science club members for their exemplary grades in science classes and their sole extracurricular attention to science and school service. These handy students, Meister recalled, "became masters of the school environment." They applied their scientific knowledge to repairing plumbing and electrical problems in the school, operated audiovisual equipment, and even implemented new technological features such as class bells and an intercom system. "Grade B" science club members enjoyed fewer privileges, but still attended meetings, lectures, and field trips. In this respect, Meister combined scientific instruction and inquiry with a civic duty to the school community. Establishing a club constitution oriented youth to parliamentary procedure, a common purpose, and allegiance to a constructive organization. If the teacher established these parameters deliberately, then students could direct their agendas with relatively little interference. Laboratory projects allowed club members to construct and manipulate "scientific toys." School-wide "science magic" demonstrations and charity work for the local community helped integrate the science club in the social life of the school. Such experiences provided not only meaningful scientific inquiry but also developed valuable civic qualities in students: aware and informed, technically skilled, and socially responsible.[26]

After nearly a decade of analyzing and initiating student projects and science clubs, Meister concluded by the mid-1920s that a widespread youth movement was feasible. Like many of his contemporary school reformers, he expressed concern about growing leisure time and the apparent rise of juvenile delinquency. Meister therefore saw science clubs as serving a vital social and civic function in a democracy: steering youth away from destructive habits and heightening their sense of community membership. He favored science clubs that carefully selected and evaluated their members, but he oriented some of their activities to the overall welfare of the school. Furthermore, Meister was convinced that the project method as applied to the science club facilitated "purposeful activity which encourages originality and inventiveness, and habituates boys to the experimental procedure." By allowing them firsthand experiences with natural phenomena and tools, students became familiar with laboratory techniques. They could "fashion raw materials into usable things" and, more generally, become equipped to assume "control of the physical and chemical elements in our environment."[27] His primary concern was not to search for and train future scientific leaders. Rather, Meister envisioned a larger purpose—one that emulated Dewey's—in considering the scientific habit of mind to be especially vital in strengthening citizens' abilities and inclinations to identify and solve problems of mutual interest. In the late 1910s and 1920s, science clubs emerged in various secondary schools across the nation with many of these objectives.

In some respects, science clubs for American youth had already been in existence for several decades. Educators in predominantly rural communities and regions had initiated out-of-school programs to teach new agricultural methods to boys and household management to girls. Contests for growing corn efficiently also became popular. The United States Department of Agriculture's organization of "corn clubs" for crop diversification in southern states complemented these initiatives. The General Education Board also began sponsoring canning and poultry clubs for girls in 1910. These various activities involved tens of thousands of rural adolescents and coalesced in the 4-H Club network with the federal passage of the Smith-Lever Act in 1914. As an agency "in training for rural citizenship," participants in 4-H clubs regularly demonstrated their projects in livestock tending, food cultivation, preparation and preservation, and farm beautification to their local communities and at state agricultural fairs. The United States' military involvement in World War I in 1917 oriented 4-H clubs to food production and preservation to secure victory. Club membership burgeoned rapidly—from 169,000 youth in 1916 to more than half a million by 1918. More than five million rural boys and girls participated from 1915 to 1924. Overall, 4-H clubs aimed to encourage and equip rural youth for a productive life of farming. Their activities typically took place beyond the physical confines of local schools. Schoolteachers—and specifically science educators—do not appear to have had much direct involvement. Instead, agricultural extension agents spearheaded these programs.[28]

By contrast, the types of science clubs that Meister envisioned tended to eschew agricultural or vocational aims, and they typically functioned in schools under the supervision of science teachers. Some clubs sought to develop greater student interest, understanding, and achievement in their science courses.[29] Others were created to cultivate constructive hobbies for students' leisure time and to promote worthy social habits such as school service and nature conservation.[30] A nature study club at Phillips High School in Birmingham, Alabama, for example, aimed to acquaint members with natural laws and to see themselves "as part of this great scheme, with certain powers and privileges and consequently certain duties and responsibilities."[31] Members of the Bird Boosters at Bowen High School in Chicago made it their mission to study birds for the purpose of protecting them and supporting the Audubon Society.[32] The science club at Upper Darby Senior High School in Pennsylvania, meanwhile, required members to serve the school by maintaining and operating projectors, a public address system, and a photography lab.[33] Clubs sometimes combined these goals to promote rational thinking and experimental skills "to discourage false and superstitious beliefs of living things."[34]

Science club activities typically included lectures, laboratory experiments, school assemblies, social events, and field trips to museums, factories, and nature sites. For instance, students in the science club at Broughton High School in Raleigh, North Carolina, regularly visited the National Academy of Science and Smithsonian Institution in Washington, D.C. At the Irwin Ave. Junior High School in Pittsburgh, students in the Edisonian Science Club prepared for lectures by finding the answers to questions submitted by fellow members about that particular subject. Furthermore, most clubs adopted constitutions and parliamentary procedures to enable students to conduct their meetings without too much teacher involvement. As Meister had recommended, they aimed to develop desirable social and organizational skills.[35]

Science clubs frequently exhibited many of the purposes and characteristics pioneered and advocated by Meister, but nothing resembling a coordinated movement would appear until the late 1920s and early 1930s. Although a majority of American adolescents attended secondary schools by the 1920s, very few took courses in general science, biology, chemistry, or physics.[36] 4-H clubs had proliferated for several decades but were largely based on farms and homes to improve agricultural living and production. By contrast, Meister envisioned science clubs aligned with the school science curriculum. In this spirit, clubs in various midwestern and southern states began to form as Junior Academies of Science affiliated with the State Academies of the American Association for the Advancement of Science (AAAS). That campaign originated in Illinois.

Junior Academies of Science

In the early 1910s, a few secondary school educators in Illinois began to incorporate the project method in their science teaching. Several scientists at the University of Illinois, meanwhile, began to encourage high schools in the state to develop science clubs. By the end of the decade, the secretary of the Illinois State Academy of Science, J. L. Pricer, advocated the development of high school science clubs to spur greater student interest in the field. Pricer urged his colleagues to recognize the importance of extending the academy's mission beyond research: "Every member of the academy should have some interest in the problems of science education in the secondary schools." Five clubs joined the State Academy in the 1919–1920 school year. A committee of scientists and science teachers also formed a High School Club Section to attract more teachers and students.[37] At its meeting in 1923, George W. Hunter of Knox College recounted his experiences as a biological field club advisor at DeWitt Clinton High School in New York City, which, as historian Philip Pauly has shown, was one of the pioneering sites of the biology curriculum.[38] According to Hunter, science clubs simultaneously strengthened student–teacher relations, served the school community, and led students beyond the laboratory on field trips. Despite these endorsements, however, relatively few teachers and students affiliated with the Illinois State Academy in the early to mid-1920s. Many of its professional scientists openly doubted that high schools could teach science well or that students could engage in serious scientific work. Perhaps for these reasons, secondary school teachers remained somewhat aloof.[39]

Nonetheless, it was the initiative of a science teacher, S. Aleta McEvoy, that catalyzed the science club movement in Illinois. In 1928, McEvoy accompanied one of her students from Rockford High School to the State Academy meeting, where the student presented his chemistry project to the High School Science Section. The academy's secretary, Lyell J. Thomas, appointed McEvoy as chair for the following year's program, and he invited the state's biology and chemistry teachers' associations in the hopes of garnering greater student participation and improving science teaching in the high schools. Thomas appointed an organizational committee, comprising McEvoy, along with Rosalie M. Parr, of the State High School Chemistry Teachers' Association, and Louis Astell, a science teacher at West Chicago High School, which spawned the Junior Academy of Science. Thomas also extended invitations to colleges and normal schools in the hopes of grooming future science teachers and sustaining the State Academy's membership. Student and teacher interest grew quickly. In 1929, 35 club sponsors accompanied 200 student delegates to the State Academy meeting in Macomb, where they exchanged information about club activities and heard scientists speak about new professional opportunities. This momentum carried into the 1930 meeting at the University of Illinois, where 175 high school students exhibited 130 of their own projects in physics, chemistry, and biology, and competed for prizes. In that year, 22 high school clubs joined the State Academy of Science. By involving some of the state's high school science teachers, Thomas, McEvoy, and their colleagues sought to foster collaborations among professional scientists, public educators, and high school students.[40]

Leaders of the Junior Academy in Illinois described their initiatives at AAAS meetings. At the December 1929 conference in Des Moines, Iowa, Astell called on other states to develop their own Junior Academies by sponsoring science clubs. Like Morris Meister, he proposed that programs beyond the school curriculum best accommodated John Dewey's pedagogical recommendations for active and socially oriented student learning. The prestige of state academies, Astell believed, could allow students "to sense the power that science is wielding." Through lectures, reprints of scholarly research, and professional guidance, experts in various states could stimulate "scientific interest and endeavor which the student will not discontinue when he leaves the high school." Astell considered teacher cooperation and development to be critical in this regard and recommended the implementation of teacher training courses and conferences. As high school enrollments had reached unprecedented heights, he wanted science clubs to influence the "hordes of children who did not possess a background of scholastic interest." These remarks echoed many educators' concerns that the rapidly rising numbers of first-generation high school students, largely immigrant and working class, were ill suited for rigorous academic work. For Astell, science clubs extended the school's supervision of the leisure time of adolescents. He viewed them as "preventative measure[s]" against juvenile delinquency and one of "the most effective ways of breaking up youthful gangs." At the same time, he believed that many students could benefit from advanced work in science. A nationally coordinated network of high school science clubs could produce a broadcasted lecture service and help them affiliate with professional scientific organizations.[41]

News of the Illinois Junior Academy of Science spread swiftly. In Kansas, for example, George E. Johnson, who had attended the 1929 meeting in Des Moines, and Hazel Branch, a faculty member at the University of Wichita, founded a junior academy in the following school year. As its inaugural president, Branch argued that Junior Academies of Science were timely in an era of rapid social and political change: "Monarchies are falling, governments are reorganizing, [and] old conditions are breaking off everywhere." Junior academies would motivate high school students to begin preparing for scientific careers at an earlier age. Branch also articulated a larger civic purpose. Boys and girls would learn "that science knows no political boundaries; that all nations and races are working toward the same end: the education of the masses in the ways of scientific thought, that through understanding all misunderstanding will pass away."[42] By the spring of 1932, 54 students from four high school science clubs had joined the academy in Kansas.[43]

These developments pleased Otis Caldwell, one of the nation's leading science educators and director of the Institute of School Experimentation at Teachers College. In 1920, Caldwell had led the Committee on the Reorganization of Secondary Education in recommending a science curriculum that reflected students' home, school, and community life, and for all high schools to offer courses in biology, chemistry, and physics. As general secretary of the AAAS and chair of its Committee on the Place of Science in Education, Caldwell hoped to initiate a national science club movement by facilitating communication among various Junior State Academies. "Science must have a better way to get down to the lower levels, to get hold of those who may be able to contribute to scientific advancement," he urged at the AAAS meeting in December 1931. In addition to engaging students in productive leisure activities, science clubs affiliated with state academies of science needed to locate and support "young students who have capacity for later doing productive scientific work." Caldwell had long contended that science education must cultivate popular attitudes open to "thinking, judging, and acting" and inform public policies for adapting to new technology and discoveries. His convictions reflected the science education and popularization forays of the AAAS in the interwar decades including lecture tours and radio programs, although a dearth of financial resources limited their development and impact. Nonetheless, the number of states with Junior Academies of Science by 1931 included Illinois, Indiana, Iowa, Kansas, North Carolina, Ohio, Oklahoma, Tennessee, Texas, and West Virginia. These organizations sponsored science clubs with activities including lectures, field trips, and annual meetings where students exhibited their projects.[44]

To accelerate this trend, Caldwell enlisted the assistance of Astell, whose flurry of lectures and articles had made him a national authority on high school science clubs. Sharing Caldwell's and Meister's conviction that a "scientific habit of mind" was critical for societal progress, Astell exhorted professional educators to lead the way: "It is time that we, as science teachers, were doing the constructive things within our power to prevent the origin of further legal barriers to scientific progress." Indeed, popular understanding and appreciation of scientific methods appeared to be sorely lacking in American society. Astell's reference to "legal barriers" may have been about the Scopes trial of 1925 that stemmed from state legislation prohibiting the teaching of evolutionary theory in schools. He pointed to teacher-led after-school programs as the remedy: "Science clubs can contribute a sympathy for, and an understanding of, science with its manifold benefits for civilization."[45] As professional science organizations including the AAAS and the National Academy of Science had struggled to foster widespread understanding and appreciation of science, Astell may have turned to schoolteachers as a potentially influential constituency.[46]

In 1932, Astell disseminated the results from his national survey of 150 science clubs. They encompassed a wide combination of subjects including astronomy, biology, chemistry, general science, geology, physics, aeronautics, photography, film, radio, x-ray, and nature study. Club activities included lectures by student members and outside speakers about their research, reports on field trips to various sites, talks about famous scientists and current developments in the field, and live experiments. Astell attributed their popularity to the concurrent rise of supervised activities for high school students in the 1920s and early 1930s, and he contended that clubs affiliated with Junior Academies of Science could combat juvenile crime and gang behavior. Science clubs developed social and leadership skills as part of character building. Equally important, they deepened students' knowledge of facts and applications, and more broadly, a systematic way of thinking and learning to appreciate science.[47]

By the early 1930s, a host of purposes had informed the proliferation and coordination of science clubs and Junior Academies of Science. Some suggested, as Morris Meister had argued in New York, that clubs fostered meaningful student learning and encouraged scientific ways of thinking to empower active citizens in a democracy. At the same time, Meister, Astell, and others emphasized that science clubs, as part of a larger program of carefully supervised student activities, could thwart juvenile delinquency. Still others hoped that the development of Junior Academies of Science could serve a vocational function by heightening high school students' interest in science and the possibilities of pursuing scientific careers. All of these aspirations fueled the development of science clubs affiliated with the AAAS. A new movement in science education had begun.

Science Fairs

High school students began to display their science club projects at annual meetings of various state academies of science in the late 1920s and early 1930s. It was during this time when the idea of a full-fledged science fair solely for youth emerged in New York City, whose State Academy of Science had not established a junior division. These public exhibitions had distinct institutional origins and purposes.

Science fairs for youth started in 1928 as part of a new educational mission of the American Institute of the City of New York, a century-old organization initially chartered by the New York State Legislature "for the purpose of encouraging and promoting domestic industry in this State and the United States."[48] The American Institute began to showcase innovations in agricultural machinery at annual fairs in New York City in the late 1820s and 1830s. By the mid nineteenth century, its industrial fairs featured a wide range of inventions such as the Morse telegraph, McCormick reaper, Singer sewing machine, Bell telephone, and Remington typewriter. The American Institute also used its exclusive membership of business and political leaders to urge public policymakers to invest in internal improvements.[49] By the end of the nineteenth century, however, its annual expositions were less profitable, as specialized industrial shows had become the norm. In many respects, the organization's original mission had been fulfilled, as the United States was emerging as a leading industrial power. The American Institute thus lacked a clear purpose in the early twentieth century. It suffered a pronounced decline in membership, and consequently, lost much of its influence and stature.[50]

A series of failed attempts to regain that prominence in the mid-1920s included sponsoring an exposition for inventors and even resuming agricultural fairs.[51] In 1927, however, two leading members with scientific affiliations charted a new course. H. H. Sheldon, professor of Physics at New York University and science editor of the *New York Herald Tribune*, and L. W. Hutchins, a public relations executive who had led a botanical expedition in the Canadian Rockies, promoted a distinct mission: "To focus the attention of the industrial public on science and scientific research, and to explain to the intelligent public the current achievements of science."[52] The

American Institute's role would effectively be that of an intermediary and advocate: "Science, particularly fundamental scientific research, needs popular presentation that the general public may better understand its importance and in turn be led to support it."[53] This purpose spawned a series of ambitious programs. Sheldon and Hutchins led a new Committee on Activities in the spring of 1927 that hosted ten luncheons for businessmen and scientists to surmise industrial applications of "pure" scientific inquiry.[54] Furthermore, to secure the interest of the public in an era when entertaining leisure activities such as motion pictures, radio, and sports were becoming increasingly popular, this committee rented theater facilities to dramatize discoveries from research laboratories. In the spring of 1928, a commemorative dinner and a book, *A Century of Industrial Progress*, celebrated the American Institute's centennial anniversary.[55]

Hutchins and Sheldon acknowledged that these new activities would be considerably costly.[56] Yet these programs increased the American Institute's membership, and significantly, more scientists and science educators joined its ranks. It was in this institutional context that the trustees voted on June 20, 1928 to allow Hutchins to initiate a new "worth while show."[57] With this broad charge, he began to plan for a children's fair to be held at the American Museum of Natural History in October. In collaboration with the museum, the School Nature League, and New York City's schools, the event intended to foster students' appreciation of nature and conservation. Believing that a fair complemented the American Institute's interpretation and promotion of science, Hutchins envisioned that it would emphasize "work in the interest of scientific farming and the biological sciences in so far as they bear on agriculture, country life and living, and local nature."[58]

This goal reflected curricular developments in the city's high schools. As historian Philip J. Pauly has demonstrated, biology courses aimed to encourage students to develop their analytical thinking and apply their knowledge to pressing problems in urban living, including nutrition and hygiene, physical health, and sanitation.[59] Historian Sally Gregory Kohlstedt has similarly explained the enduring popularity of the nature study movement in New York City that offered a plethora of "large public parks, natural history museums, botanical gardens...[and]...zoos" as educational resources.[60] Press releases for the children's fair announced that it would show teachers, students, and other visitors the value of school gardening and nature work already occurring in the city's schools. Planners hoped that "conservation of our natural resources may be understood and appreciated" by all.[61]

The School Nature League worked closely with American Institute officials in planning for the Children's Fair. Founded in 1917 by Alice Rich Northrup, a naturalist who turned her attention to education, the organization furnished nature study rooms in schools and held seasonal flower shows. Northrup and her colleagues believed that students' and teachers' improved knowledge and appreciation of their natural environment would make them better citizens. For city residents, this meant recognizing the relationships between nature and public health. School gardens and nature rooms would also teach youth the necessity of collaboration in addressing problems of mutual interest. When Marjorie Coit assumed the School Nature League's directorship in 1928, the American Institute found a ready partner to collaborate with the local school board. Coit secured exhibits from numerous youth and conservation organizations, including the Brooklyn Botanical Society, the National Plant and Flowers and Fruit Association, the Brooklyn Children's Museum, the Boy Scouts, Campfire Girls, and the New York City's Department of Parks. In addition, Van Evrie Kilpatrick, head of the School Gardens Association, selected and arranged displays for the Children's Fair.[62]

An elaborate classification system for fair exhibits emerged. The first division was for groups including school science clubs, parks, and other youth organizations. It consisted of seven categories: gardens displaying fruits, flowers, and vegetables; conservation exhibits (forests and flowers, birds and animals, or parks and roadsides); biological principles featuring adaptation and evolution; plants and conditions for their growth; artificial selection for improvement in plants; "economic crops"; and plant and animal life for classroom use. The second division, designated for individual boys and girls, also consisted of seven subject categories. Three of these—gardens, conservation, and biological principles—mirrored those in the group division. In addition, individuals could submit projects on the life history of an insect or prepare a cage of living insects; construct a cage for larger animals; make plaster casts of animal tracks, tree leaves, or seedlings; or submit "the best and most original notebook or record book" for a particular living organism. Cash prizes would "be used for gardening, nature study, conservation equipment or books." This organizational rubric reveals the heavy emphasis on nature study and the biological sciences. Subjects in chemistry, the physical sciences, and engineering were notably absent. The rapid rise of biology as a school subject in the early twentieth century, the enduring popularity of school gardens and nature study in New York City's schools, and the School Nature League's close involvement oriented the fair in this direction.[63]

The inaugural Children's Fair, held from October 18 to 21, 1928, in the Education Hall of the American Museum of Natural History, proved to be immensely popular. On the eve of the fair's opening, Coit announced that more exhibits had been submitted than could be displayed. Nearly 3,000 local youth presented their science projects, and more than 35,000 students, teachers, and city residents attended. Judges awarded 103 cash prizes to participants. The site of the fair was especially prestigious, as natural history museums were widely seen as institutions of scientific and cultural authority in American society. In New York City, the American Museum of Natural History was an especially influential educational institution. In a given year during the 1920s, an estimated six million children visited the museum.[64]

The emphasis on agriculture and nature study was prevalent at the Children's Fair. The nature "notebook or record book" category received the greatest number of entries. Special "feature displays" from the Bureau of Children's Farms and Demonstration Gardens included two models: "One, of a city park as it should look, clean and spotless; the other, littered with papers and refuse, as the parks frequently are left by careless visitors." Other exhibits intended to demonstrate the virtues of "good roads" and "forest conservation."[65] Students from Newtown Agricultural High School exhibited "chickens, a swarm of bees and a model of the best type of farms, farm implements and farm buildings."[66] The fair also featured lectures by science educators including Paul B. Mann, head of the Biology Department at Evander Childs High School, who discussed the "Conservation of Wild Life" and "Birds and Their Relation to Man." Curators from the American Museum of Natural History, as well as agricultural scientists, also delivered addresses. Albert Russell Mann, dean of Cornell University's College of Agriculture, encouraged his audience to appreciate the importance of developing students' conceptual understanding of their natural surroundings: "If we proceed to analyze logically, step by step, starting with simple plants or animals and studying their structure and their processes, moving forward to more difficult forms and functions, we pass into the realm of what we call science."[67]

Delighted by its popularity, the American Institute's trustees resolved to make the Children's Fair an annual event. President Edwin F. Murdock gushed about the widespread "commendation and praise" his organization received. Echoing some science educators and the leaders of the nature study and school gardens movements, he articulated the science fair's civic dimensions: "We should remember that these children, whom we have interested in a subject so pertinent to good citizenship, are the coming generation." Meanwhile, some teachers and students thanked the American Institute for organizing the fair. A teacher from the Bronx expressed her appreciation of "the interest that it aroused in the nature work of our schools." The principal of an elementary school in Manhattan reported that "Miles Horek, winner of an individual prize magnanimously presented his award to our Nature Room to be used with the School prize for the benefit of our Nature Room." Reflecting on the American Institute's active programs for 1928 including science lectures, the centennial dinner, the commemorative book, and a new in-house publication, Murdock communicated that students were already planning projects for the next science fair to be held in the fall of 1929.[68]

The American Institute's leaders nonetheless scrutinized the financial costs of their new programs. In 1928, for instance, the organization's expenses outweighed its income by a staggering $17,352.[69] As a result, Murdock led a movement among the trustees in the summer of 1929 to suspend these activities until they could become financially self-sustaining. Some senior officials, meanwhile, objected that the emphasis on science enervated the American Institute's agricultural roots. The trustees subsequently dismissed Hutchins from his position as director of activities, and the fate of the Children's Fair became uncertain. This institutional retrenchment prompted a contentious struggle within the organization. By the summer of 1930, a large contingent of members with scientific affiliations prevailed. Murdock was removed as president and succeeded by A. Cressy Morrison, fellow of the New York Academy of Sciences and president of the Union Carbide and Carbon Company. Otis Caldwell, who had been promoting the education programs in State Academies of Science, was elected to the vice presidency. Despite the economic crisis of the Great Depression, this new leadership assured that the science fairs would continue each year as a central component of the American Institute's established mission.[70]

The Children's Fair became increasingly popular, and its stated aim persisted: "To focus attention on the sciences and to foster a scientific interest in agriculture, gardening, nature study and conservation."[71] An array of organizations—the National Association of Audubon Societies, the Society for the Prevention of Cruelty to Animals, the Wildflower Preservation Society, the Garden Club of America, and Newtown Agricultural High School—prepared special exhibits in 1929. On "Boy Scout Day" at the fair, members of Nature Study Troop 72 presented a statue of a scout "symbolizing the 'Nature Study Scout Spirit'" to School Nature League officials. The nature study and life sciences orientation therefore remained prominent. At the same time, new members of the American Institute's planning committee introduced entry categories in the physical and chemical sciences. The central figure in this development was none other than Morris Meister.[72]

Meister proposed new exhibit categories that derived from his research and experience with scientific toys and student workrooms. He recommended a competition for elementary school age youth with "the best display of inventiveness with science toys" that illustrated a physical law or involved a mechanical model "utilizing Meccano parts, Erector parts, Tinker Toy parts." For older children, Meister envisioned prizes for the best equipment, models, or illustrations of "adaptations and applications of scientific principles for the problem of living in a modern city home." Similarly, he devised entry categories for adolescents who constructed "displays of homemade useful commodities, such as soaps, dyes, candles, inks, metal polish, reflectors, fire extinguishers, paper" with samples of the original ingredients.[73]

Many aspects of the 1929 Children's Fair exemplified Meister's educational priorities. New entry categories for "Inventiveness in the home," "Chemistry in the home," and "Models illustrating physical principles" all followed his proposals. Astronomy, geology in nature, and geology in man-made structures joined the individual entry categories as well. Meanwhile, three categories from the inaugural fair—nature notebooks/record books, conservation, and backyard gardens—were discontinued. Meister's innovations reflected his convictions that meaningful student learning derived from actual life problems. They familiarized youth with experimentation and methods of controlling chemical and physical phenomena. New generations of citizens could then apply those skills to improving their communities. Hutchins informed the high school science teachers of New York City that entry categories in physics, chemistry, and biology were implemented "especially to interest high school students." These new categories appeared to be appealing; models illustrating physical principles comprised the greatest number of entries (60) among the 283 entered in the individual class.[74] In the coming years, the subject emphasis at science fairs would continue to shift toward the physical and chemical sciences. Emblematic of some science educators' criticisms of nature study as sentimental and insufficiently rigorous, these changes gradually discouraged younger children from participating in the fair.[75]

Meister also recommended holding the fair later in the school year to allow teachers more time to guide their students' projects. Van

Evrie Kilpatrick, the longtime advocate of school gardens, recognized the implications: "[We] must decide if [the] Fair is to be an agricultural, horticultural, or gardening Fair or one of natural sciences." As crops and gardens could not last for exhibits in the winter or spring, those types of projects would be precluded. "If natural science is the dominant motif," Kilpatrick declared, "it should be in one of the proper weeks."[76] School Nature League representatives subsequently agreed to move the date to early December and acknowledged that agricultural exhibits would need to be omitted.[77] The quest to allow students and teachers more time to develop projects of higher quality, coupled with an attempt to draw more secondary school participants, superseded the science fair's initial focus on nature study and conservation.

Pushing for a greater emphasis on the physical and chemical sciences, the fair's planning committee began to call on the city's high school science teachers to encourage their students to prepare exhibits. Nationally, the status of physics and chemistry courses was somewhat mixed. Although actual student enrollments in these subjects rose from 1915 to 1934, the proportion of public high school students enrolled in physics fell from 14 to six percent. The proportion taking chemistry held steady at seven percent during this period.[78] At the Children's Fair, new subject categories for 1930 and 1931 embodied a dizzying array of 63 different entry classes for individuals and groups at the elementary, junior high, senior high, and nonschool divisions. For elementary school-age youth, the vast majority of subjects remained oriented to conservation and the life sciences. By contrast, most of the additions to the junior and senior high school divisions—"chemistry," "principles of physics," and "the best models or apparatus to show how certain useful inventions work"—reflected a new curricular focus on the physical, chemical, and mechanical sciences. At the same time, other categories including "industry, mining, and farming" for young children, "transportation and communication" for early adolescents, and "science in home and city life" for high school youth belonged to ongoing efforts to cultivate students' awareness of how science shaped their immediate surroundings.[79]

Fair organizers also began to publicize the science fair more widely. Speaking on the WNYC radio station in December 1930, Coit touted the School Nature League for furnishing nature rooms in the city's schools, managing the Children's Fair, and encouraging youth to appreciate the importance of conservation.[80] Caldwell, meanwhile, pointed to the fair as clear evidence that children were spending their leisure time productively: "An excellent place to take a carping adult who 'feels so sorry' about our younger generation." The fair helped youth to decipher the complexities of the modern city. According to Caldwell, innumerable manifestations of science and technology abounded: "When one looks over the apartment buildings of New York or of any other large city anywhere in the civilized world, a wilderness of radio antennae presents itself...As one walks the streets or through the parks he sees toy airplanes that will fly, and toy boats that will sail." The Children's Fair therefore played a vital role in maintaining and directing the interest and ingenuity of youth to productive social ends: "In back yards, vacant lots, and in the homes are thousands of illustrations of children's experiments and of their inventiveness...of their everlasting curiosity in doing things with the natural phenomena about them." Familiar examples illustrated science's relevance and the need to understand its applications to urban life. "Subways, skyscrapers, milk and water supplies, good air and good food, freedom for play and to be healthy, these and other city needs compel us to understand and to properly use the science of our day," Caldwell warned, "else we pay the age of old penalties that have always been exacted by ignorance and carelessness." Like his former colleague at Columbia University, John Dewey, and former student, Morris Meister, Caldwell contended that students' curiosity of their immediate surroundings motivated their scientific understanding. The Children's Fair would encourage local youth to direct that thirst for knowledge to tangible ends for improving urban communities.[81]

Promotional literature similarly articulated the fair's social and civic benefits. In a book featuring selected student exhibits, Coit argued that the scientific principles behind modern living must be understood. Coit acknowledged that "it might be possible for many of us to live in the midst of the marvels of nature and of man and, closing our minds to why and how, to press a button or turn a switch and find the miracles of modern civilization appear and disappear." "But how meager an existence this would be!" she exclaimed.[82] As evidence that the fair helped students and teachers to recognize the impact of scientific research on daily life, Coit featured student projects from nature study programs. A sixth grade teacher from Brooklyn described contrasting scenes in her students' exhibit on "The Value and Protection of Forests": one depicting harmony among careful lumbering, farming, and the forest and the other portraying indiscriminate clearing and a dilapidated shack.[83] According to a junior high school teacher from Queens, her students' fair project on seed dispersal led to "searching in fields, woods, and gardens" and "put into action our class motto, Agassiz's 'Study Nature, not books.'"[84] A junior high school student, who had displayed 100 insect specimens collected from the city, traced his project to an early childhood love of nature.[85]

Coit began to suffer from a prolonged illness in 1931, and in her absence, the School Nature League's involvement in the Children's Fair ended. The American Institute assumed most responsibilities, including working directly with the schools and selecting exhibits. Meister renamed it the Children's Science Fair and explained that the new title "confirm[ed] our belief in the Fair as a potent force in science education." The quest to facilitate firsthand observation and experimentation to deepen students' understanding of scientific phenomena and appreciation of investigative methods persisted. According to Hugo Newman, principal of New York's Teacher Training College, active and socially conscious citizens would emerge. Newman praised the fair on NBC's *Great Moments in Science* radio program on November 4, 1931, for teaching children how to investigate their natural environment and enlisting "the scientific habit of mind" to strengthen their communities.[86]

By the early 1930s, financial deficits stemming from the Great Depression compelled American Institute officials to curtail some of their programs. They nonetheless determined to sustain the science fairs, and Meister's influence within the organization continued to grow. He joined its Board of Trustees and Board of Managers, and chaired the fair planning committee. The science fair in 1931 remained popular; 176 groups and 229 individual students displayed their projects. Meister's publication of a book detailing the fair's short history, moreover, aimed to inspire similar initiatives in other communities. Indeed, educators from other parts of the nation were beginning to notice. A representative from the Buffalo Museum of Science and a faculty member at Western Reserve University in Cleveland each visited the 1931 fair to gather ideas for initiating their own.[87]

As an advocate for a more comprehensive emphasis on science education in all school grades, Meister lamented the "meager" amount in elementary schools that was often limited to nature study.[88] Nationally, about 18 percent of elementary schools taught no science in the early 1930s, and only 20 percent furnished a classroom devoted to science instruction.[89] Meister also protested gender segregation in the junior high school curriculum and too great an emphasis on biology over other subjects. If scientific concepts only became meaningful when derived from students' experiences, he argued, then New York City's schools should place less emphasis on nature study and biology: "The present practice tends to ignore the real experiences available in an urban environment." "To continue the emphasis upon purely biological sciences in the elementary junior high school and later grades," he cautioned, "is to entrench a vicious circle into which a better rounded curriculum in science can never enter."[90]

The Children's Science Fair could remedy this apparent curricular deficiency. Students who prepared projects for much of the school year gained enduring and valuable firsthand experiences in science. "Whether it be the chief purpose of science teaching to interpret environment, build generalizations or inculcate thinking habits," Meister explained, "participation in the Fair provides the raw materials out of which any of these aims are possible of achievement." To encourage fair visitors' active consideration of the civic consequences of the science projects on display, Meister prepared and distributed "guide sheets" for elementary, junior high, and senior high school age youth. These posed questions about natural resource conservation and related social problems. Regarding an exhibit titled, "Contrast of Sanitary and Unsanitary Streets," for example, elementary school students were asked, "Is it more fun to live on a sanitary street? Why?" Questions about the "Parks" exhibit included: "Are you a good citizen? How can you keep your park like the pretty ones here?" Junior high school students were urged to consider how they could improve the aesthetic quality of their homes and to anticipate "how shall we travel when all the coal and oil have been used?" Senior high school students, meanwhile, should ponder the importance of family planning: "What is the relation between number of offspring and parental care?" In these ways, the Children's Science Fair oriented students' projects to larger societal problems and possible solutions.[91]

In the late 1910s, science clubs had originated to furnish meaningful learning experiences consistent with the new project method. Advocates including Morris Meister, Louis Astell, and Otis Caldwell argued that clubs could cultivate rational and empathetic citizens in a participatory democracy. Junior Academies of Science also acquainted some high school students with professional scientists and possible career opportunities. In New York City, the American Institute's Children's Science Fair aimed to cultivate children's awareness of nature and conservation in an urban environment. With the eventual decline of nature study, these fairs began to embrace a wider array of scientific subjects, particularly in the chemical and physical sciences. Meister engineered this transformation—one that reflected larger curricular changes in American science education. Within a few years, the fairs had become an annual fixture for the city's schools. Emboldened by this popularity, American Institute officials, led by the untiring Meister, would begin to enlist the city's science teachers to form an elaborate network of science clubs. A science education movement for civic ends now seemed within reach.

CHAPTER 2

Building a Network

With their initial success, organizers of the annual Children's Science Fair in New York City began to think more ambitiously about their role in science education. The American Institute's leaders also searched for ways to increase membership and augment their financial security, particularly in light of growing expenses and declining assets resulting from the Great Depression. Across the United States, more than 5,000 banks failed between 1929 and 1933, and the Gross National Product declined by half in that span. While public schools struggled to sustain their operations and curtailed their social services, teenagers found themselves increasingly excluded from the job market. High school enrollments rose as a result, but many youth also began to enjoy more leisure time, which led some educators to worry about the prospects of juvenile delinquency. Like the vast majority of American communities, New York City and its schools suffered enormously from the economic woes of the early 1930s. By 1933, roughly one quarter of the city's workers were without jobs. In some areas, such as manufacturing and mechanical trades, unemployment rates exceeded one-third. The municipal government bordered on bankruptcy, and as a result, several thousand teachers lost their jobs from 1930 to 1931, while others endured salary reductions. After-school, evening, and summer programs that had aimed to extend the school's influence in the lives of youth and their families were eliminated. At the same time, the city's high school enrollments ballooned 45 percent from 1930 to 1935, which swelled class sizes and crowded school buildings.[1]

As these conditions hindered hands-on and individualized experiences in science classes, the rationale for expanding the American Institute's science education programs intensified. A committee chaired by Morris Meister began in 1931 "to look into the advantages of some type of connection between the American Institute and the High School Clubs and the formation of Junior Membership." Building on the American Institute's history of promoting industry and forging relationships with professional scientists, laboratories, and manufacturing plants, the group devised an elaborate plan for science club activities. Meister and his colleagues proposed programs led by adult experts to facilitate students' scientific investigations and to acquaint them with scientists and their research. Such efforts could compensate for overcrowded schools and increased leisure time by encouraging active inquiry and critical thinking: qualities that a participatory democracy required of its citizens.[2]

In May 1932, Meister detailed a preliminary plan to the city's school principals, heads of high school science departments, and science club teachers. He pointed to the annual Children's Science Fair, and the American Association for the Advancement of Science's (AAAS) Junior Academies, as evidence of considerable student interest. The American Institute could encourage greater student participation, Meister argued, by summoning the city's intellectual and cultural resources and furnishing prizes for worthy accomplishments. Students would visit scientific laboratories, local industries, and museums. Monthly "science expositions" in schools would feature club members' projects, while demonstration lectures "by eminent men of science" would serve "as a stimulant to the pupils." An annual "Junior Science Clubs Congress," moreover, would emulate AAAS meetings with large audiences witnessing student experiments and hearing leading scientists speak about their areas of expertise. Meister's committee thus envisioned a citywide network of science education activities.[3]

Various American Institute officials endorsed this plan. Treasurer Alfred Knight asserted that a series of experimental demonstrations for youth beyond regular school hours could compensate for parents' apparent ignorance of science. Paul B. Mann, chairman of the Biology Department of Evander Childs High School and associate at the American Museum of Natural History, believed that an expanded organization could help to locate and groom future scientific leaders. These programs could also provide students valuable experiences unavailable in schools: contacts with scientists and their research institutions. Not all favored the increased expenditures that these programs would incur, however. Most notably, former Treasurer H. T. Newcomb worried about the American Institute's financial instability in an era of economic decline. Knight responded by vigorously defending the Junior Activities: "My heart is in this projected work, as I do not believe there is any plan which, if carried out properly, could have more far reaching results than that which has been so carefully worked out by Dr. Meister." He also argued that an initial investment in the science education programs would strengthen the American Institute's prospects for philanthropic donations. Privately, Knight disparaged Newcomb's reservations as antiquated, and he praised the American Institute's emerging leadership of scientists and science educators.[4]

With this support, and the cooperation of the city's board of education and science teachers, Meister's group implemented the new programs in the 1932–1933 school year "to encourage, stimulate and guide science club activities in and among schools to the end that science may become more integrated in the life of the pupils."[5] On a local radio broadcast, Meister proposed that the science fairs and clubs could compensate for "the deadening atmosphere brought about by parents who dismiss child interests with gestures of impatience." As the Great Depression had exacerbated crowded classrooms and hindered curricular innovation in science, moreover, they would support "over-burdened teachers" and provide opportunities for students to pursue interdisciplinary projects related to their experiences living in an urban environment. "The program of instruction in the schools, unaided by the contributions from extra-curriculum activities," Meister explained, "cannot possibly interpret the place of science in modern life."[6] He was convinced that all students "must learn to think like a scientist" to solve everyday problems and "to live properly in this modern age."[7] From Meister's perspective, the American Institute's new Junior Science Clubs would play an indispensable role in civic education.[8]

New Programs

Thousands of students from New York City joined the Junior Science Clubs in the mid-1930s. They participated in demonstration lectures, museum and workshop courses, Science Congresses and Christmas Lectures, and the annual Children's Science Fair. To help coordinate these activities, the American Institute inaugurated a bi-monthly newsletter for science teachers and students, *The March of Science*. Issues recommended fruitful projects and experiments, reported on club activities and field trips, identified scientists volunteering to meet with clubs, posted laboratory equipment for exchange, and announced interclub events. The American Institute required each affiliate to pay dues of one dollar per semester; 117 clubs joined by the spring of 1933.[9]

On the last Saturday morning of October 1932, H. H. Sheldon welcomed 657 Junior Science Club members to their first interclub meeting at the American Museum of Natural History. Students heard Oscar Riddle of the Carnegie Institution speak about the future of biological research, and City College Professor Ross A. Baker discussed "Chemistry Today and Tomorrow." John A. Clark, chairman of the Physics Department at Alexander Hamilton High School, dispensed advice about conducting experiments in physics. Students also heard Meister delineate "the Junior Science Clubs Plan." Sheldon, Meister, and the event's organizers hoped to supplement students' current studies and to "equip them with an understanding of the possibilities of scientific achievement in the future."[10]

"General demonstration lectures" featured scientists conducting experiments for student audiences. In the spring of 1933, Meister and his colleagues sent information letters and ticket order forms to over 300 secondary school principals, high school teachers chairing science departments, and student editors of school newspapers. They also issued a press release to 142 publication outlets. On March 25, more than 1,200 student members of the Junior Science Clubs attended one of the two museum sites. Those visiting the Museum of Science and Industry heard staff member Robert P. Shaw tell "the Story of Electricity." Shaw aimed to engage the students by depicting "a series of pedagogic exhibits" dramatizing "the classical experiments of famous scientists of the past." Four short films in the museum's theater depicted various electrical principles and applications. At the American Museum of Natural History, Paul B. Mann described "The Museum's Part in Exploration" and in educating millions of Americans about life on Earth. Films documented the collection of dinosaur eggs in the Mongolian Desert and how to mount fossil vertebrates. Curators then escorted students to one of the museum's five halls to focus on a particular subject in the life sciences.[11] As historian Steven Conn has shown, natural history and science museums increasingly sought to educate youth as a distinct audience over the course of the twentieth century. In New York City, the museums' collaborations with the American Institute exemplified that trend.[12]

Demonstration lectures would continue to attract thousands of club members. On November 4, 1933, for instance, students attended sessions in junior science, senior biology, or senior physical science at either the College of the City of New York or the New York Botanical Gardens. Raymond L. Ditmars, curator of reptiles and mammals at the New York Zoological Park, lectured on "Strange Animal Friends," while Lincoln T. Work, a professor of Engineering at Columbia

University, discussed "the Significance of Fine Particles in Chemical Engineering." In the fall of 1934, a national broadcast of the "Radio Explorers Club" at the NBC studios featured sea captain James P. Barker, who detailed his journeys around Cape Horn. James P. Clark, vice director of the American Museum of Natural History, shared his expertise on the rhinoceros. Club members met with both speakers after the broadcast. At a demonstration meeting in the spring of 1935, Kenneth Blanchard, a biology professor at New York University (NYU), explained how to determine molecular structures, while Wanda K. Farr, of the US Department of Agriculture, described her experiments on the fiber of cotton plants. American Institute officials hoped that these presentations would "lift science above the routine of the classroom" and demonstrate "its place in their own lives" as students contemplated pursuing comparable projects.[13]

A new series of "museum" or "workshop" courses similarly drew hundreds of science club members as well as dozens of professional scientists, museum curators, and science teachers. A course on aviation for high school students in the spring of 1934 consisted of three meetings at the New York Museum of Science and Industry, where two-dozen students witnessed demonstrations of aerodynamics, motors, and instruments used in test flying. Junior high school students in the course on mineralogy, meanwhile, attended nine sessions at the Brooklyn Children's Museum, where they learned how to examine and classify minerals.[14] Subsequent workshop courses offered senior and junior high school club members the chance to study applied subjects, including railroads, electricity, "nature handicraft," and "technique of making habitat groups."[15]

The American Institute partnered with NYU in developing workshop courses in biology and chemistry for high school club members as well. On seven Saturday afternoons in a campus laboratory, 20 students interacted with different faculty on biological subjects including "micrurgy," cytology, biochemistry and microbiology, endocrinology, physiology, genetics, and biophysics. The inaugural chemistry course, by contrast, enrolled 400 students, who attended nine weekday afternoon lectures on the history of chemistry, microchemical analysis, colloids, physical chemical measurements, x-rays, electrochemistry, spectroscopy, organic chemistry, and toxicology. According to the American Institute's director, L. W. Hutchins, the workshops acquainted "young scientists" with investigative methods and knowledge beyond the city's overcrowded schools. "These courses at New York University will be conducted like college science work," he explained, "but they will be perfectly intelligible to secondary school students having a familiarity with chemistry and biology."[16] Subsequent workshop courses in chemistry would be limited to 20 students "so that members may have the opportunity of performing experiments in college chemistry under special guidance."[17] New courses in physics and aeronautics joined the array of offerings, as well as one for teachers about experimental techniques in a classroom laboratory. At the American Museum of Natural History, 15 junior high school students could enroll in a workshop course on "nature handicraft," while 12 high school club members could register for a course on "mineralogy" at the Brooklyn Children's Museum.[18]

In addition to these programs, an annual Science Congress featured dozens of selected demonstrations by students and professional scientists. At the inaugural event in May 1933, nearly 1,000 Junior Science Club members attended one of eight section meetings featuring student experiments at the American Museum of Natural History. In all, 53 students representing 23 science clubs conducted these sessions. Alfred A. Berger of Franklin K. Lane High School in Brooklyn, for instance, demonstrated "How To Prepare and Make Microscope Slides," while Miriam Gold and Martha Berman of the Girls Commercial High School performed "Experiments With a Bunsen Flame." Organizers had met with science club sponsors and students weeks in advance to determine which projects were most worthy of inclusion. Designed to emulate professional scientific meetings such as the AAAS, the students also fielded questions from their peers. In the afternoon, the entire student audience assembled to watch the Principal of Stuyvesant High School, Ernest von Nardroff, demonstrate "The Physics of Sand."[19]

The Science Congress proved to be a popular annual gathering. In 1934, for instance, the event featured 89 students leading 12 sessions on subjects including organic chemistry, airplanes, and vacuum tubes. At the 1938 Congress, students addressed topics from the fields of biology, chemistry, physics, astronomy, photography, and "science hobbies." In addition to having the opportunity to demonstrate scientific principles for several hundred of their peers, a few students received special recognition from corporate sponsors. In 1936, the General Electric Company selected one student with the best presentation for a visit to its research laboratory in Schenectady and to deliver the talk as part of its *Science Forum* radio program. Two years later, the Carnegie Institution of Washington awarded one trip to its fish hatcheries at Cold Spring Harbor, while the Westinghouse Electric and Manufacturing Company hosted an outstanding student speaker at its factories and research laboratories in Pittsburgh.[20]

Although it elicited widespread interest, far more boys than girls participated in the Science Congress during its first six years. In 1933, for example, girls comprised only nine of the 53 students (17 percent) presenting their experiments. Three girls, all from Grover Cleveland High School, participated in the biology group in demonstrating aspects of "Frog Raising" and "Ant Raising." Three of the seven students in the chemistry group were girls. However, no girls were among the 26 boys who presented experiments in general physics, applied physics, photography and light, and physical science for younger students. These patterns persisted in the coming years. Only 47 of the 340 students (14 percent) who demonstrated their science projects or laboratory techniques at the Science Congress from 1933 to 1938 were girls. Female students remained especially underrepresented in particular subjects. Of the 185 students presenting in the physical sciences, only one was a girl: Edith Schreiber from Brooklyn Children's Museum who measured the thickness of rock strata. Gender inequalities in the biological sciences were also evident but less severe; girls comprised 22 of the 88 (25 percent) students presenting. In no year did the overall percentage of girls demonstrating experiments at the Science Congress exceed 21 percent. As historian Kim Tolley has shown, the rise of domestic science and commercial courses in secondary schools during the early twentieth century, coupled with the declining percentage of girls taking advanced mathematics courses, contributed to corresponding declines in chemistry and physics. It appears likely that these curricular and social trends similarly limited the participation of girls conducting experiments at the Science Congress.[21]

Meister did not acknowledge these gender disparities in touting the American Institute's larger initiatives in science education. Through open forums in the morning section meetings, he argued, club members from different schools could learn about each other's work and find an "outlet for expression and criticism." Teachers sponsoring science clubs, meanwhile, could exchange ideas about programmatic and organizational matters. Meister contended that participants gained inspiration from emulating professional organizations like the AAAS, and they learned that "the true scientist does not hide his discoveries from the world. He seeks full and free discussion by his colleagues." Professional scientists, according to Meister, "make their annual pilgrimage to a central meeting place in order to present their findings, check their thinking and receive further impetus in the search for truth." Yet the congress's primary purpose was not to train future scientists. Welcoming the 1,500 students attending in 1934, Meister proposed that science clubs benefited society in a more general way. Without the burden of classroom bells, homework, and exams, students could select "life-long hobbies" and acquire habits "for the most satisfying use of future leisure time." Regardless of students' vocational aspirations, all could develop rational thought through constructive science activities. In an era when many Americans worried that high youth unemployment would foster rebellion against adult authority or legitimate radical political ideologies, educators like Meister emphasized the social and civic virtues of science clubs and congresses.[22]

The popularity of demonstration lectures, workshop courses, and science congresses prompted the American Institute to add annual Christmas Lectures resembling those of the Royal Institution of London. Held over two days in December 1934 at the American Museum of Natural History, the inaugural event featured biologist Robert Chambers and chemist and Nobel Laureate Harold C. Urey. It also included Jean Piccard, who spoke about his explorations of the stratosphere, and *The New York Times* correspondent Russell Owen, who described an expedition to the South Pole. With 1,500 students in attendance, these talks were broadcast across the nation through the NBC radio network. According to Hutchins, the Christmas Lectures compensated for "overcrowded conditions" in schools and enlivened "routinized methods" of teaching science. Inviting "junior science clubs over the country to take advantage of the opportunity of listening in," he hoped that students and educators beyond New York City would appreciate and emulate the American Institute's initiatives. In 1936, Harlow Shapley, director of the Harvard College Observatory, simulated how the sun would appear from a remote vantage point in the universe as "an Astronomical Explorer Broadcasting from Antares." G. Edward Pendray, a founder of the American Rocket Society and science writer, discussed the possibilities of interplanetary exploration. Both Shapley and Pendray would play important roles in organized science activities for youth in the coming years.[23]

American Institute officials continually touted the benefits of their educational programs in the hopes of fostering a widespread youth movement in science. Speaking to a neighborhood association in New York City, Hutchins profiled a female science club member, aspiring to become a biologist, who had benefited from the support of her parents and science club sponsor in conducting successful chemical experiments on blood. According to Hutchins, the student gained further confidence from the positive response she received after presenting her results at the Science Congress: "Her work was given so much recognition that

Figure 2.1 The American Institute's Science Congress program cover from 1939. Collection of the New-York Historical Society.

one of the big scientific instrument makers heard of her and presented her with the latest type [of] microscope used for studying the composition of blood." The student subsequently enrolled in one of the biology workshop courses, which Hutchins believed would "help bridge the gap between her high school and her college work."[24]

At an AAAS meeting on "the Place of Science in Education" in December 1933, Meister outlined the American Institute's educational programs to an audience of science teachers. He lamented educators' perennial predicament: unable to placate supervisors, prepare students for examinations, and accommodate students' interests and queries because of the rush to cover the curriculum. By contrast, after-school science clubs offered more flexibility and "appeal[ed] to the individualized interests and the stimulus to creativeness." Meister also hoped that these initiatives would spread: "That science teachers' associations undertake the task of coordinating the science clubs in their respective localities, along the lines developed by The American Institute in New York City." Despite the rigid constraints of established curricula in the regular science classroom, "the club s pirit...[could]...break down the sharp barriers between school and after-school." Teachers sponsoring clubs, moreover, became popular with students, enlisted more varied pedagogical methods, and fostered students' critical thinking through laboratory experiments. These essential practices, according to Meister, developed inductive reasoning that equipped youth "for enriched living in the modern world of science and in a democratic form of government."[25]

Indeed, the architects of these science education programs often asserted that all participating youth would benefit socially and civically—regardless of whether they aspired to scientific careers. In December 1937, the American Institute's new director, Gerald Wendt, and junior activities coordinator, Catherine Emig, touted the recently completed Science Congress on their local radio series, *Accent on Science*. Wendt described the meeting as an open forum of knowledgeable and "earnest students" engaging in "perfectly free discussion...of the facts and what they mean" and without "vanity or appeals to the emotions"—all of which simulated "the democracy of the future." He recounted that a student presenter, who had painted a clock to glow in depicting the effects of radium, confessed to an audience member's challenge that the object did not contain the radioactive element. According to Wendt, this exchange highlighted both the "integrity" of the student demonstrator, who chose to respond honestly, and the keen skepticism of students in the audience, who were "more interested in the truth than in the show." Those in attendance learned that "no real scientist can be a bluffer" and "science also means accepting the conclusions whether you like them or not." Emig, meanwhile, contended that the American Institute's educational programs "are of most value to the boys and girls who are not going to be scientists later but will become lawyers and journalists and business men and housewives." "That honesty, that detached point of view, that habit of facing the facts and what they mean," Emig concluded, "is certainly a habit that will benefit anybody." By encouraging critical thinking and the free exchange of ideas, science clubs taught valuable lessons to citizens in a democracy.[26]

In addition to these civic justifications, the American Institute's leaders sought to increase membership, fortify finances, and extend their influence beyond New York City. An alliance with the American Chemical Society's Division of Chemical Education in 1933 meant that all club members also belonged to its Student Science Clubs of America—a loose conglomeration of subscribers to the weekly magazine *The Science Leaflet*. American Institute officials envisioned an expanding network of science clubs across the United States. Despite the financial cost, they hoped to attract new generations of paying members. Some of the short-term gains were impressive. By the 1935–1936 school year, 5,907 students in 232 science clubs, nearly all from New York City secondary schools, affiliated with the American Institute.[27]

Some ventures did not fare as well. The American Institute's Junior Activities Committee attempted to assist the Crime Prevention Bureau of the New York City Police Department. In 1933, Police Commissioner Edward P. Mulrooney invited Hutchins to establish science clubs in various precincts: "To reclaim youngsters by interesting them in science."[28] Fearing that the closing of summer schools and disrepair of playgrounds stemming from the economic depression would lead to idleness and delinquent behavior, Mulrooney asked the American Institute to train selected police officers to become science club sponsors. Despite Meister's initial hesitancy, he prepared and led a training course for 50 police officers.[29]

Yet this partnership proved to be unrealistic. Some doubted that impoverished or delinquent children would show interest or could benefit. There was also the matter of finding qualified adults to lead the clubs. Many of the officers who enrolled in Meister's training course found it to be "too academic" and expressed considerable difficulty in comprehending the material. In response, some proposed supplying the officers with books from the public library to guide activities in less demanding subjects including "auto mechanics," "how to make puppets," and "how to breed fish." Meister objected for fear that it would compromise "science" by introducing "crafts." The results of this social experiment were therefore decidedly modest. Activities in the summer of 1933 were reduced to hiking trips led by police officers and possibly accompanied by a "trained scientist...who will interest as many of the children as he can in nature study...[and]...geology." It appeared that the American Institute's science education programs were best suited for students with an interest and inclination in science by having voluntarily joined a club. This fleeting collaboration with the Crime Prevention Bureau highlights some of the limitations of the American Institute's rhetoric about the civic benefits of science clubs. Despite its lofty visions, only a small fraction of high school students from New York City and across the nation participated in science clubs in the 1930s. Without proficient teachers who believed that all youth were capable of understanding science, it seemed that only a select segment of American youth could acquire the skills and dispositions for democratic citizenship.[30]

The Children's Science Fair

Amid the proliferation of new activities, the Children's Science Fair remained the American Institute's most prominent educational program. A press release from 1936 described a diverse array of exhibits in the Education Hall of the American Museum of Natural History, which resembled "a zoo, a laboratory, an industrial plant and a hangar." Visitors experienced auditory stimulation with "the roar of a wind tunnel" and "the music of a model gramophone." They could gaze at models of an "interplanetary rocket ship, a blast furnace, a green-headed, ring-necked Dinosaur, a skyscraper in the process of building." According to fair organizers, the results of scientific inquiry could be both enlightening and entertaining. More generally, as American museums competed with new leisure activities in the interwar decades, some public displays of science aimed to dazzle audiences with the material products of research.[31]

Fair organizers also hoped to convince parents that their children's scientific projects required their support: "That these domestically disturbing factors may command a new respect and find encouragement instead of deprecation at home." Belonging to "a modern, progressive movement in science education...of national importance," the fair inspired students "to put into actual practice his class room observations" and "encourage originality of thought." American Institute officials also claimed that the annual event was attracting the attention of educators throughout the world "as a most valuable method of encouraging the study of science in secondary schools." As a result, audiences should appreciate "the significance of this movement for education, for science, and for the future of our civilization."[32]

Hundreds of student exhibitors and tens of thousands of visitors attended the science fair each year. Exhibits belonged to one of ten subject classifications, and their relative popularity shifted over time. In 1932, exhibits in the plant and animal life category were most prevalent with 95 of the 478 on display (19.9 percent). The biology division was the second most popular with 74 (15.5 percent) of student exhibits. Combined with the 28 conservation and 34 health entries, projects from the life sciences amounted to 231 of the 478 student exhibits on display (48.3 percent). In the four succeeding science fairs, however, the physics division garnered the greatest number of exhibits. Whereas physics had the third highest number of entries at the 1932 fair with 71 (14.8 percent), within a year it became the most popular with 113 of the 523 total exhibits entered (21.6 percent). Physics retained its dominant position in the coming years. While the entries in physics generally increased, and those in biology held relatively steady, there was a significant drop in the number of entries in the plant and animal life division in this period, and the health subject division was eliminated after 1933. Table 2.1 illustrates the number and proportional representation of projects in the entry categories for the five science fairs held between 1932 and 1937.[33]

As with the Science Congress, more boys than girls participated in science fairs and received prizes for their exhibits. This gender disparity occurred in all subject areas, albeit in varying degrees, for both junior and senior high school students. Table 2.2 depicts the distribution of students' exhibits by aggregated subjects from grades 7, 8, and 9 for the seven science fairs held from 1930 to 1937. Girls produced only 120 of the 771 junior high school student projects on display (15.6 percent). In the life sciences categories—including biology, plant and animal life, health, and conservation—girls entered 69 of the 255 projects (27 percent). Girls were especially underrepresented in the physical sciences. Female students prepared only six of the 44 exhibits in astronomy (13.6 percent). In physics and chemistry, this disparity was even more pronounced. Girls displayed only one of the 32 junior high school-aged exhibits in chemistry (3.1 percent) and just 10 of the 213 entries in physics (4.7 percent). In the geological sciences (consisting of the earth studies category), girls' representation was relatively higher, with 13 of the 44 student exhibits on display (29.5 percent). In the miscellaneous grouping—consisting of history of science, "industries," and "energy"—exhibits prepared by girls amounted to only 21 of the 183 total (11.5 percent). Females in junior high school were thus significantly underrepresented at the American Institute's science fairs.[34]

Similarly aggregated data reveal the same pattern at the senior high school level. Table 2.3 depicts the distribution of students' exhibits by grouped subjects for grades 10, 11, and 12. Girls contributed only 143 of the 755 student exhibits on display for the seven fairs held during these years (18.9 percent). The life sciences were most popular among

Table 2.1 Children's Science Fair projects by subject, 1932–1937

Subject	1932	1933	1935	1936	1937	Total
Stars & the Solar System	23 (4.8%)	22 (4.2%)	29 (6.1%)	21 (4.0%)	15 (3.2%)	110 (4.5%)
Earth Studies	22 (4.6%)	27 (5.2%)	34 (7.1%)	46 (8.8%)	39 (8.5%)	168 (6.8%)
Plant & Animal Life	95 (20.0%)	82 (15.7%)	63 (13.2%)	67 (12.8%)	55 (11.9%)	362 (14.7%)
Biology	74 (15.5%)	82 (15.7%)	70 (14.6%)	79 (15.0%)	72 (15.6%)	377 (15.3%)
Physics	71 (14.9%)	113 (21.6%)	111 (23.2%)	152 (29.0%)	110 (23.9%)	557 (22.6%)
Chemistry	48 (10.1%)	57 (10.9%)	43 (9.0%)	37 (7.0%)	31 (6.7%)	216 (8.8%)
Health	34 (7.1%)	32 (6.1%)	0 (0.0%)	0 (0.0%)	0 (0.0%)	66 (2.7%)
Conservation	28 (5.9%)	17 (3.2%)	27 (5.6%)	36 (6.9%)	29 (6.3%)	137 (5.6%)
Industries	64 (13.4%)	72 (13.8%)	52 (10.9%)	48 (9.1%)	67 (14.5%)	303 (12.3%)
History of Science	17 (3.6%)	19 (3.6%)	37 (7.7%)	32 (6.1%)	30 (6.5%)	135 (5.5%)
Energy ("science idea")	0 (0.0%)	0 (0.0%)	12 (2.5%)	7 (1.3%)	13 (2.8%)	32 (1.3%)
Total	476 (100%)	523 (100%)	478 (100%)	525 (100%)	461 (100%)	2,463 (100%)

Sources: "The Fifth American Institute Children's Science Fair," 1932, NYHS AIR, Box 145, Folder 3, 9; "The Sixth American Institute Children's Science Fair," 1933, NYHS AIR, Box 150, Folder 6, 5; "The Seventh American Institute Children's Science Fair," [1935], NYHS AIR, Box 150, Folder 6, 9; "The Eighth American Institute Children's Science Fair," 1936, NYHS AIR, Box 183, Folder 1, 12; [No Title], 1937, NYHS AIR, Box 193, Folder 12. Courtesy of the New-York Historical Society.

Table 2.2 Seventh, eighth, and ninth grade boys' and girls' exhibits at the Children's Science Fair, 1930–1937

Subject grouping	Boys' exhibits	Girls' exhibits	Total exhibits
Astronomical (stars and solar system)	38 (86.4%)	6 (13.6%)	44 (5.7%)
Biological (plant & animal life, biology, and health and conservation)	186 (72.9%)	69 (27.0%)	255 (33.1%)
Chemical (chemistry)	31 (96.9%)	1 (3.1%)	32 (4.2%)
Geological (earth studies)	31 (70.4%)	13 (29.5%)	44 (5.7%)
Physical (physics)	203 (95.3%)	10 (4.7%)	213 (27.6%)
Miscellaneous (history of science, industries, and energy)	162 (88.5%)	21 (11.5%)	183 (23.7%)
Total	651 (84.4%)	120 (15.6%)	771 (100%)

Source: "Children's Science Fair, 1930–1937: Distribution of Exhibits," NYHS AIR, Box 209, Folder 14. Courtesy of the New-York Historical Society.

Table 2.3 Tenth, eleventh, and twelfth grade boys' and girls' exhibits at the Children's Science Fair, 1930–1937

Subject grouping	Boys' exhibits	Girls' exhibits	Total exhibits
Astronomical (stars and solar system)	24 (92.3%)	2 (7.7%)	26 (3.4%)
Biological (plant and animal life, biology, and health and conservation)	225 (74.0%)	79 (26.0%)	304 (40.3%)
Chemical (chemistry)	109 (87.9%)	15 (12.1%)	124 (16.4%)
Geological (earth studies)	45 (78.9%)	12 (21.0%)	57 (7.6%)
Physical (physics)	140 (90.9%)	14 (9.1%)	154 (20.4%)
Miscellaneous (history of science, industries, and energy)	69 (76.7%)	21 (23.3%)	90 (11.9%)
Total	612 (81.1%)	143 (18.9%)	755 (100%)

Source: "Children's Science Fair, 1930–1937: Distribution of Exhibits," NYHS AIR, Box 209, Folder 14. Courtesy of the New-York Historical Society.

senior high school students, with 304 (40.3 percent) of all exhibits. Girls prepared 79 (26 percent) of these projects, which approximated the gender ratio in the junior high school grades. Even fewer female high school students were represented in the physical sciences. Girls presented only two of the 26 exhibits in astronomy (7.6 percent). The ratios in chemistry and physics were comparable. Female students prepared 15 of the 124 projects in chemistry (10.8 percent). In physics, girls constructed 14 of the 154 projects exhibited (8.3

percent). These figures were slightly higher than the gender ratios at the junior high school level. In the geological sciences (consisting of the earth studies category), girls' projects comprised 12 of the 57 on display (21.1 percent). In the miscellaneous grouping, consisting of history of science, industries, and energy, exhibits prepared by girls amounted to only 21 of the 90 total (23.3 percent). Taken together, girls prepared only 263 of the 1,526 exhibits (17.2 percent) displayed by junior and senior high school students at the seven science fairs held from 1930 to 1937.[35]

These trends reflected broader developments in American science education. As historian Kim Tolley has demonstrated, curricular and pedagogical shifts in the late nineteenth and early twentieth centuries began to alienate many elementary and secondary school girls from scientific study. The rise of home economics, clerical, and other gender-distinct vocational curricula, moreover, drew some girls away from the life and physical sciences. The decline of nature study in elementary schools and a new emphasis on textbook instruction may have diminished girls' enthusiasm for science as well. The portrayal of the physical sciences as masculine disciplines similarly contributed to declining female enrollments in high school chemistry and physics.[36] Furthermore, historian Paula Fass's analysis of students' extracurricular choices at seven New York City high schools from 1931 to 1947 demonstrates comparable patterns. Fass found that 50 boys belonged to physics clubs, while only 7 girls did so; 84 boys joined chemistry clubs, compared to 64 girls. At the same time, nearly twice the number of girls belonged to "other science" clubs than boys: 218 compared to 110. The gender disparity in chemistry club participation was not as wide as the gap in science fair participation from 1930 to 1937. Despite girls' overwhelming majority in the "other science" clubs, they were severely underrepresented at the annual science fairs.[37]

Although fewer girls than boys participated in New York City's science fairs each year, some fair organizers and science club sponsors actively encouraged and promoted girls' involvement. Catherine Emig proclaimed that "some of the most interesting projects are those done by girls." In 1932, she praised Lillian Mayer's exhibit, "The Development of Electric Motors," Dorothy Orline's model of celestial objects, "The Universe in an Umbrella," and a junior high school girl's depiction of New York's water supply system. A press release describing student achievements at the 1936 fair similarly highlighted girls' scientific accomplishments, two of whom had won first prizes. The exhibits, both from girls attending Haaren High School in the

Bronx, featured a model of a Ford V-8 engine built from an old sewing machine and curtain rods, and a model steam engine.[38]

Some exhibits demonstrating the applications of science to daily living, however, revealed gendered differences. Emig claimed that "boys have a more marked tendency toward industrial subjects and girls toward science in the home." She also cited the prominence of girls' projects from the 1932 fair on subjects such as "The Care of Babies," "Chemistry in the Home," "First Aid," and "Medicine." It did not surprise Emig that "no boy laid a finger" on exhibits such as these, because they "added the application of this world of science to the problems peculiar to those spheres of life ordinarily presumed to be a woman's."[39] In other words, if gender differences manifested themselves in the occupational and social realms, then similar distinctions should emerge from a science pedagogy that championed experiential learning. Emig seemed to value highly the "domestic" applications of science, because they helped to make science meaningful to girls. At the same time, such messages may have taught some female students that their future roles—whether as homemakers or scientists—were to be confined to domestic matters. As Tolley has explained, the constrained job market during the Great Depression "fortified cultural assumptions about the secondary status of women in the workplace." Curricular recommendations and reforms, moreover, attempted to align the teaching of science "to students' presumed social needs and interests." These tendencies, coupled with outright job discrimination in the sciences and colleges of education, often conspired to discourage many girls from envisioning scientific careers.[40]

Fair organizers also hoped to draw and hold the attention of thousands of spectators. As a result, judging criteria at the science fairs rewarded students not only for originality, but also for a polished artifact. Agnes G. Kelly, staff assistant at the American Museum of Natural History, instructed prospective exhibitors to envision a shop window display: "It demands Beauty or one of Beauty's associates in some form or another." "It demands attraction through color, arrangement, the bizarre, or the unusual," Kelly explained: "It must attract the wandering eye and satisfy the restless mind after the magnet has played its part."[41] To an audience of teachers in 1935, Hutchins similarly touted the virtues of advertising criteria in the public presentation of scientific research: "The exhibit fails of its purpose if it doesn't tell a story to the observer clearly and quickly."[42] Emig, meanwhile, sought to ensure that all of the fair exhibits were engaging. "This is a *show*," she emphasized, "the visitors are to be taken into consideration."[43]

The evaluation rubric issued to the science fair's judges each year reflected these criteria, including "clearness of objective," "importance of the idea," "accuracy," "originality," "workmanship," "effectiveness of the presentation of material," and "general attractiveness, neatness, and care."[44] A project's quality of presentation weighed as heavily as its scientific accuracy and originality. A successful exhibit should be "arresting," "clean and well finished," "well illuminated," and have an "aesthetic appeal." "Free from unnecessary wordiness," its purpose should be "clearly conveyed."[45] In other words, scientific creativity and ingenuity were insufficient: students needed to communicate the central idea of their research clearly and to present it in such a way that riveted the attention of onlookers. In light of these considerations, some fair organizers regularly lamented the prevalence of "technical exhibits that are not easily comprehensible to the general run of students and other visitors."[46]

This scrutiny may have arisen from concerns about declining attendance and increasing operational expenses. The fair in 1933 had a peak of 41,019 visitors over six days. Cooperation from the Board of Education of the City of New York, which encouraged principals and teachers to allow students to attend during school hours, may have helped to funnel attendance. As Meister and the members of the Junior Activities Executive Committee began to plan for the science fair in 1934, however, Hutchins indicated that some of the American Institute's leadership believed that holding the fair was financially unsound. Expenditures for the Science Congress, Christmas Lectures, demonstration workshops, and newsletters in 1934 had nearly tripled the amount spent on youth programs three years earlier. These factors likely informed the decision to cancel the fair in that year.[47] It was reinstated in 1935, but attendance continued to decline: from 31,430 in 1935, to 23,280 in 1936, to 21,358 in 1937.[48] Although school enrollments had reached new heights in the early years of the Great Depression, the number of students attending schools in New York City declined by approximately 40 percent from the mid-1930s to the mid-1940s. Among the city's high schools, moreover, greater curricular differentiation and ability grouping reinforced the notion that only the most academically talented youth could benefit from rigorous and enriched programs. The American Institute's quest to promote good citizenship through science education appeared to be eluding large segments of students and adults. By 1936, moreover, its trustees issued an ultimatum: The Student Science Clubs must become financially self-sustaining within the coming year or face termination.[49]

In response, the American Institute's director, Gerald Wendt, championed these programs for cultivating critical thinking and creative expression. They were especially relevant, because modern society was "based on science and in the days to come this understanding of the world will be more and more necessary."[50] Fairs and congresses, Wendt declared on CBS radio in 1937, would produce socially responsible leaders: "Scientists of the future will not be monks or hermits. They will understand the society in which they live. The statesmen of the future will not merely ask science for more machines. They will themselves think scientifically about social problems." Wendt thus envisioned science club activities for all American youth: "When this great work with New York school children is extended to all our youngsters then, using the word in the high sense as I have, the scientists of the future will be, quite simply, the next generation."[51] A student writing in the science club newsletter, *Amateur Scientist*, similarly claimed that his generation would solve the "many racial, economic, and social problems that baffle mankind today." "The method of science is the only intelligent way," the student asserted: "It includes observation, collection of facts, testing, careful judgment, and conclusion." Such an evaluation of societal affairs would preclude war or other acts of force, "the way of the savage for which there is no place in a civilized world."[52] These ideas reflected an emerging political awareness among American scientists in the 1930s. As historian Peter J. Kuznick has shown, growing numbers of professional scientists began to contend that their methods of investigating the natural world could yield rational and systematic solutions to the nation's social and economic problems. In this context, the prospects for lasting societal reform helped to persuade the American Institute's leaders to sustain their costly science clubs, fairs, congresses, and workshop courses.[53]

The prospects of favorable publicity leading to membership increases, corporate sponsorship, and wider societal influence appeared alluring as well. Indeed, these programs increasingly interested professional educators and scholars in New York City and beyond. Graduate students from Teachers College, NYU, and New York Teacher Training College visited the Children's Science Fair as part of class assignments. Representatives from the New York Principals Association, the Physics Teachers Club of New York, and the New York Association of Biology Teachers attended the fair in 1933. NYU's School of Education selected 30 exhibits from the 1935 science fair to display at its summer school for teachers. American Institute officials welcomed this attention, and they worked actively to cultivate it. They arranged for "traveling exhibits" of outstanding student projects to various sites.⁵⁴ Award-winning fair exhibits were on display at the December 1937 AAAS meeting in Indianapolis and at the National Education Association convention in 1938. This "signal honor and distinction," American Institute officials hoped, "should be a significant step in our widening field of operations."⁵⁵

Beyond New York City

While the American Institute developed its elaborate educational programs, a number of State and Junior Academies of Science affiliated with the AAAS continued to guide science clubs for secondary school students. Each year, scientists and science teachers exchanged information about their educational initiatives at the national Academy Conference. At the December 1932 meeting, Otis Caldwell urged Junior Academies to coordinate their efforts and pointed to the science club networks in New York City and Illinois as valuable precedents. With the AAAS, he had also chaired the Committee on the Place of Science in Education, which held an essay contest for high school students across the nation. Its purpose, Caldwell explained, was "to give recognition and encouragement to those young persons who possess unusual interest and capacity for constructive work of high quality."⁵⁶ He subsequently cofounded and directed the National Committee on Coordination of Junior Academies of Science, whose members included representatives from Texas, Indiana, Iowa, and Illinois. Caldwell hoped to foster a movement in science education that would identify, encourage, and prepare talented youth to become scientists. Like the American Institute, the Junior Academy of Science's leaders sought to acquaint interested students with professional scientists, compensate for inadequacies in secondary school science instruction, and cultivate productive habits for democratic citizenship. Indeed, as the economic hardships of the Great Depression burdened and disrupted families, some Americans began to fear that youth could be swayed by radical ideologies or turn to criminal activities. These sorts of concerns partly motivated the campaign to expand Junior Academies of Science in the 1930s. Not solely for professional training, these programs would orient adolescents to civic values that would prize rational thought and secure a degree of social and political stability.⁵⁷

By 1938, 13 states and one municipality had established active Junior Academies of Science with more than 300 science clubs and thousands of secondary school students. Some professional scientists applauded this development and invited student members and science club sponsors to attend their meetings. The Nebraska State Science Teachers Association, for instance, planned a joint conference in 1937 with the High School Section of the Nebraska Academy of Science. Founded in 1931, Pennsylvania's Junior Academy membership grew to roughly 1,200 students by 1937. In a gathering that emulated the American Institute's Science Congress, Pennsylvania's student officers presided over a joint session with the Senior Academy, where other students read papers and conducted experiments. Participants also attended talks from some of Pennsylvania's scientists about their subjects of expertise. To promote camaraderie among clubs, moreover, most Junior Academies issued newsletters or shared information about their activities in their states' academy of science journals.[58]

Some of these educators hoped that science club members would develop rational thought for democratic citizenship. The Academy Conference's secretary, S. W. Bilsing, proposed that state academies must ensure the further growth of clubs "to acquaint the general public with the value of science in every-day living and also to acquaint a larger group of people with the ideals of scientific thinking." Regardless of a student's future vocation, Bilsing argued, science clubs developed "the ability to think clearly and to properly evaluate the problems of life."[59] As general secretary of AAAS in 1938, Caldwell urged state academies to supplement the school curriculum, improve science teaching, and combat the prevalence of irrational thought. Such efforts were equally critical in instilling better ways of thinking: "When science clubs initiate new members by proving that breaking mirrors does not bring bad luck; by showing the initiates the wrong of 'bearing false witness' against another person; by showing the wrong of making untruthful reports...they surely are helping society."[60] These convictions belonged to what historian John C. Burnham has identified as a longstanding and increasingly difficult struggle of scientists and educators to combat the prevalence of superstition among the majority of Americans.[61]

Others believed that Junior Academies should identify and groom civic leaders. Karl F. Oerlein, who had helped establish Pennsylvania's Junior Academy, suggested that science clubs would never become universal, because "the programs and the type of work do not appeal to the slow moving group of students." As a result, academies must prepare "future intellectual and professional leaders...for responsible citizenship in an even greater America of tomorrow."[62] Howard E. Enders, dean of the School of Science at Purdue University, informed Indiana's teacher educators in 1933 that Junior Academies must identify and encourage future scientists by acquainting them with the rigors of the profession.[63] These views resembled those of many other professional educators in the 1930s who assumed that only a select portion of the growing numbers of high school students could benefit from a robust curriculum in the sciences.

Beyond State and Junior Academies, some educators across the nation sought to emulate the American Institute's programs. A science teacher at Dunbar High School in Washington, D.C., for instance, asked for resources to develop club projects and plan for a science fair. A physician at the Mayo Clinic in Rochester, Minnesota, learned about the fair from Gerald Wendt at the December 1937 AAAS meeting and solicited printed literature to share with local teachers and students. A professor of physics at the University of Arkansas requested a sample science fair program as a planning guide for a fair at neighboring high schools. The Wollaston Mothers' Club of Quincy, Massachusetts, initiated a science fair in 1936 with ten subject entry classes identical to those in New York City. Its brochure thanked the American Institute and American Museum of Natural History "for their interest and encouragement."[64]

Some educators from across the United States also visited the science fairs in New York City. The curator of Chicago's Museum of Science and Industry observed the 1932 fair in hopes of initiating his own. A faculty member of the Rhode Island School of Design searched for ideas to incorporate in its annual hobby exhibit for youth. The director and the science teacher of Elizabeth Peabody House—a settlement home for boys in Boston—visited the 1935 fair to gather ideas for holding a science fair for their students. A junior high school science teacher in Jackson, Michigan, sought to arrange for one of her students to travel to the 1938 fair to observe and "bring back as much enthusiasm to other students and also teachers." The teacher subsequently organized Jackson's first science fair.[65] By the late 1930s, moreover, some students and teachers beyond New York City became members of the American Institute. Students from Dickinson and Irvington High Schools in New Jersey, and Chaminade High School in Long Island, participated in the 1937 Science Congress. By 1938, nine groups from New Jersey schools and nine others from New York City's suburbs had joined the Student Science Clubs. Students sought membership from as far away as Detroit. American Institute officials also solicited exemplary student science projects from communities in California and Oklahoma to display in New York City.[66]

The American Institute's educational initiatives also inspired more ambitious programs in other states. From 1935 to 1937, members of the New Jersey Department of Agriculture, in conjunction with the Department of Public Instruction, held a statewide science fair. They aimed to further students' and teachers' scientific knowledge, to inspire other students to engage in scientific inquiry, and "to bring before the public the achievements of children in the fields of agriculture and natural science."[67] New Jersey's fair leaders had visited the Children's Science Fair, and they solicited the American Institute's guidance. The judges' evaluation criteria mimicked those used in New York City.[68] By contrast, however, New Jersey's fair placed a heavy emphasis on scientific applications to agricultural production. The inaugural event coincided with the state's Agricultural Week, and all but one of the seven entry classes reflected the orientation to agriculture and life sciences. Students entered exhibits in separate categories on the production and marketing of crops and livestock in the state. Those wanting to pursue projects in physics or chemistry had to demonstrate their bearing on agricultural production. Some of the student exhibits featured in the state's brochure included "the relation of sulphuric [sic] acid to agriculture," "honey—hive to consumer," and "the production of cranberries."[69] When the New Jersey Department of Agriculture canceled the fair after 1937, science teachers and club sponsors from several high schools began to participate in the New York City science fairs.[70]

In Oklahoma, meanwhile, an energetic high school science teacher led a statewide science club movement and consulted regularly with American Institute officials. Edith Force, from Woodrow Wilson High School in Tulsa, chaired the Oklahoma Academy of Science's High School Relations Committee to foster "better science teaching and more vitally enthusiastic students of science." She created an association of science students, arranged for contacts with professional scientists, and published a newsletter about club activities in Oklahoma, other Junior Academies, and the American Institute. In 1936, Force encouraged 75 high school students and teachers to display posters, animal specimens, and electrical devices at the Oklahoma Academy of Science meeting in Stillwater.[71] She had read Meister's accounts of the science fairs and congresses in New York City, and had listened to radio broadcasts of the Christmas Lectures. For their part, American Institute officials were pleased that educators from thousands of miles away were seeking their guidance. They pointed to the initiatives in Oklahoma as evidence of the "value of specific service the Institute can and does perform in serving as a model for activities in other states where there is interest in club work."[72]

This widespread emulation fueled the American Institute's confidence about its educational influence and potential for growth. "New

York Children's Science Fair is gradually becoming a leading event in the educational world," its leaders declared: "Similar fairs are now being held over the country, many of them under the direct leadership of the Institute." Despite declining local attendance in the mid-1930s, the fairs had drawn the attention of educators and students beyond New York City. In that sense, it was no longer a local event. By 1938, the American Institute's officials therefore believed that they were on the verge of expanding their influence by coordinating a national network of science club programs for American youth. It remained to be seen whether the founding civic ideals—cultivating rational thought and social responsibility for a healthy democracy—would flourish on a national scale.[73]

CHAPTER 3

Showcasing Young Scientists at the New York World's Fair

On April 30, 1939, a monthly magazine for educators and students, *Science Observer*, described the events in the Junior Science Hall at the recently opened World's Fair in New York City: "Twenty-one young scientists opened the 'show-case laboratory'...where eight hundred boys and girls will participate in the American Institute Science and Engineering clubs exhibit." The article noted that these activities were housed prominently "in the central portion of the Westinghouse building facing the singing tower of light. Exhibits built by young scientists from all parts of the country occupy glass-fronted show cases along the wall and around laboratory tables in the center of the room where students will carry on their work."[1] Organized by the American Institute—and with the support of the Westinghouse Electric and Manufacturing Company as well as the local school board—the 40 student exhibits represented the fields of astronomy, biology, chemistry, engineering, nature studies, photography, physics, and "physiography." Comprising the efforts of high school students primarily, but not exclusively, from the New York City public schools, these projects depicted various scientific phenomena including the effects of ultraviolet light on plant growth, how human ribs act during breathing, and the molecular features of calcium fluoride. The laboratory workshops, meanwhile, featured students demonstrating the principles of crystal growth, the grinding of mirrors, the winding of motors and transformers, and methods of preparing microscopic slides. There would also be an amateur radio station, a photography lab, and a number of ceremonial events publicizing the scientific achievements of these students to a national audience.[2]

The American Institute's leaders had seen the 1939–1940 New York World's Fair as a unique opportunity to showcase their elaborate science education programs to millions of visitors. They hoped that this event would prompt a national movement in science education—one that they would lead—to promote a brand of scientific literacy and greater public appreciation of the importance of science to societal progress. By securing the sponsorship of Westinghouse, one of the nation's most prominent industrial manufacturers, it appeared that the American Institute had the means to realize that goal.

Reclaiming a National Audience

As organizers in New York began in 1935 to plan for the World's Fair, the American Institute's trustees, managers, and members worried that the explicit profit-seeking motive of the fair would obscure or skew the value of science to society.[3] The American Institute's director, Gerald Wendt, invited representatives from 36 scientific organizations to meet on July 13, 1937, to address the issue. He lamented that "the word science has not once appeared at any point whatever in all the broad plans and detailed schedules of organizing of the Fair...Science should be everywhere; it promises to be nowhere."[4] Many of the scientists and science educators in the ensuing discussion acknowledged the challenges in convincing fair organizers to dedicate a building to science, because large industrial companies such as General Electric, DuPont, and Westinghouse had already begun constructing halls to showcase their own research. Acknowledging the entrepreneurial motive behind the World's Fair, this group began to explore how science could be depicted to the anticipated millions of visitors.[5]

Two ideas emerged that would eventually take shape. Albert Blakeslee, director of the Carnegie Institution at Cold Spring Harbor, recommended that the American Institute draw from its educational initiatives in selecting and displaying worthy student projects at the World's Fair. In a similar vein, Watson Davis, director of Science Service, Inc., in Washington, DC, envisioned a science workshop for students to conduct experiments in view of fairgoers. Both Blakeslee and Davis believed that such an approach could educate the larger public effectively about the contributions of scientific inquiry to daily living and societal progress. The American Institute thus established a committee chaired by one of its trustees, Hoyt D. Lufkin (also publicity director for the New Jersey branch of Westinghouse), to ascertain how to feature the science projects of youth at the World's Fair.[6] Determining that "a science exhibit must not steal the thunder of Industry," this group intended to complement the priorities of the fair's organizers. Envisioning a plan that demonstrated the industrial applications of scientific concepts to potential corporate sponsors, members searched for ways to attract visitors to a science exhibit that "thrilled and amused."[7]

By February 1938, however, the American Institute specified a purpose consistent with its decade-long forays in science education: To cultivate rational thought for democratic citizenship by highlighting the processes of scientific investigation. Activities would include the display of award-winning exhibits from the American Institute's annual science fair, a series of science shops or laboratories in action, an array of photographs taken by students, and the enactment of weekly student science meetings. In deciding to focus on student projects, American Institute officials hoped to popularize a distinct pedagogy: "A new method of progressive science education...[that]...permits boys and girls to work directly with the tools of science, and to perform scientific experiments and investigations in much the same manners as do adult scientists." Incorporating language reminiscent of John Dewey, they stressed the importance of fostering "meaningful experience[s]," developing a student's "interests and powers" toward a career or "leisure," and cultivating "a habit of thought" consistent with the "scientific method." They also sought "to crystallize the attention of parents, boys and girls, educators and laymen to the real possibilities of this type of work as a constructive force in the community." An additional motive was to publicize the American Institute's burgeoning network of high school science clubs that now stretched across the United States.[8]

These educators had envisioned a national organization of science clubs as early as 1934. The popularity of the American Institute's local science fairs, science congresses, Christmas Lectures, and workshop courses led some within the organization to believe that similar needs and interests in scientific understanding existed in communities beyond New York City. In turn, they argued that a national network would benefit all science clubs through the exchange of ideas about laboratory experiments.[9] As discussed in the previous chapter, financial constraints had prevented the American Institute from realizing these goals immediately, although they consulted with educators in various states. By 1937, its leaders outlined an extensive plan for science education programs through an elaborate bureaucracy of local and regional staff members, who would spur community involvement. Proposed activities included science fairs and congresses, zone and central meetings, the creation and maintenance of bird sanctuaries and small museums, and a shortwave radio league. American

Institute officials hoped to enlist the assistance of adults in improving the science education of youth in their respective communities and for each student to develop "a lasting interest in his particular subject as a basis for a life-long hobby or for his future profession." The matter of funding these proposed initiatives, however, remained in question. The World's Fair therefore appeared to present an opportunity for the American Institute to showcase its local initiatives in science education and to prompt a national movement.[10]

With respect to its designs for the World's Fair, the American Institute similarly needed financial assistance to cover expenses related to the rental of space, equipment, and service of the city's students and teachers. It solicited the cooperation of the Board of Education of the City of New York to provide roughly $38,000 for the hiring of teachers and transporting an estimated 2,000 student participants. The American Institute's leaders offered to grant the superintendent the authority to assign and oversee a planning committee composed of local science teachers. They also attempted to reassure school officials that "in no sense is this a commercial undertaking. The Institute guarantees that there will be no advertising of a commercial nature, or even suggestions of commercialism surrounding the program."[11] Despite this appeal, the school board refused to assume any financial responsibility for a science exhibit at the World's Fair.[12]

This rejection prompted American Institute officials to pursue other avenues of sponsorship more urgently, and they appealed to large business corporations. In particular, they considered a unique collaboration with one of the nation's leading industrial manufacturers: The Westinghouse Electric and Manufacturing Company. With the announcement in December 1937 that Westinghouse would be constructing a building at the World's Fair to feature its research and consumer products, Lufkin brokered an arrangement that promised to be mutually beneficial. By providing space for the American Institute to feature its science club and fair programs, Westinghouse would enjoy favorable publicity. "Westinghouse will not expect to sell a toaster or refrigerator," Lufkin argued, "but will sell Westinghouse to the rising generation and do the whole thing in a broad educational way."[13] In addition, the corporation would impress scientifically proficient students who could become future employees by "start[ing] with the youngsters at high school age and bring[ing] them up in the Westinghouse tradition."[14] With this financial support, meanwhile, the American Institute could highlight the particular contributions of science education to society as well as spawn a national network of science clubs, fairs, and congresses. Lufkin succeeded in convincing

American Institute officials—including Gerald Wendt, Robert Pollock, Paul Mann, and Morris Meister—that Westinghouse's funding would neither unduly emphasize the commercial applications of science nor compromise the democratic aims of their science education initiatives. These men also believed that Westinghouse's sponsorship for several years could yield a robust program on a national scale that could secure longer-term investments from educational foundations.[15]

For these reasons, they quickly sketched a plan to persuade Westinghouse's leaders that the American Institute was well positioned to foster a national extracurricular program in science education. Mann initially articulated a rationale that was civic in its orientation: "To foster the spirit of science in children and in adolescents to the end of producing a new generation whose scientific talents not only are discovered for the world, but whose thinking and action are accurate, based on evidence, free of prejudice, and therefore scientific and wholesome."[16] In other words, more widespread scientific understanding could foster a participatory democracy with rational citizens. Wendt and Pollock, meanwhile, claimed that tens of thousands of better-trained science students would come to feel indebted to Westinghouse, and more generally, "the very influential educational world will realize equally the public spirit of the Westinghouse Company."[17] With the World's Fair as a platform for showcasing their science education programs, the American Institute's leaders believed that an elaborate national network would be launched. This plan included regional offices with fieldworkers who would inaugurate new local science fairs and congresses and a national monthly magazine written and edited by students for club members. Organizational booklets, traveling exhibits (including a miniature model of the World's Fair), and a staff of consultants would also help to guide science club activities in various communities. Secondary schools would be targeted, but the cooperation of various youth organizations, museums, and the American Association for the Advancement of Science (AAAS) would be enlisted as well. A nationwide system of affiliated science clubs would ultimately create constructive leisure activities for American youth who would come to appreciate the contributions of scientific inquiry and knowledge to societal progress.[18]

These designs convinced Westinghouse officials to sponsor the American Institute's science education initiatives for a minimum of three years, starting in October 1938. Westinghouse agreed to donate space and equipment in its building at the World's Fair. It also consented to funding the national expansion of science fairs, clubs, and congresses. This support reduced the American Institute's budget deficit by 48 percent within the first year, which emboldened its leaders' ambition to expand their network of science clubs and fairs nationwide. They began by publishing a monthly student newspaper, *Science Observer*, and mailed the inaugural issue in December 1938 to 15,000 high school students across the United States.[19]

The American Institute announced these developments through radio programs and national publications. One of these included the AAAS's flagship journal, *Science*, in which Junior Activities Director Charles Federer Jr. explained that "through the medium of traveling organizers, a speakers bureau and eventually short-wave radio, the work of institute clubs will be organized as one national unit."[20] By November 1938, the American Institute began distributing a pamphlet, "How to Organize a Science Club," to high school teachers and principals across the nation. It claimed that science clubs engaged youth constructively and furthered human progress through active problem solving. The document presented guidelines for organizing clubs, proposed different types of activities, and offered strategies for securing experimental equipment. Its preface, authored by Meister, referenced science clubs as indicators of a "Science Youth Movement" in which "future chemists, engineers and doctors find time to explore their probable life-work and hobbyists lay the basis for joyful use of future leisure."[21] An article in the January 1939 issue of *Science Observer*, moreover, promised individual and societal benefits for joining: "Such a movement enables you young scientists to win public recognition for your achievements, to keep in touch with the work of other individuals." Eventually, those gaining distinction would "wield significant influence to preserve democracy and to retain the inherent American principle of recognizing the initiative of the individual."[22] Firsthand scientific investigation, the American Institute's leaders claimed, would assist aspiring young scientists professionally and empower millions of others as active and critically thinking citizens. Traveling consultants also visited national youth organizations' headquarters, and they hoped to subsume science clubs in various states already affiliated with state academies of science. In these ways, the American Institute actively solicited club membership beyond New York City.[23]

The immediate results of these initial efforts were mixed. Some communities began to emulate the American Institute's precedents. For example, educators and community leaders in Buffalo, New York, initiated a science congress in the spring of 1939, while Syracuse, New

York, established a science center for youth. One of the American Institute's traveling representatives also helped to organize science clubs in Western Pennsylvania. Inquiries were even received from science educators as far away as Argentina. Despite thousands of mailings, however, only two science clubs—from Bloomfield, New Jersey, and Sioux City, Iowa—joined the American Institute in November 1938. Several dozen more were added over the next three months, and on the eve of the opening of the World's Fair in April 1939, there were 84 science clubs from 27 states that belonged to the American Institute's organization, although most were from New York City. Traveling representatives appear to have reached relatively few communities as well. To fuel the proliferation of science clubs, fairs, and congresses in all corners of the United States, it appeared that the American Institute had to capitalize fully on its exhibits at the World's Fair.[24]

In announcing Westinghouse's generous financial commitment, the American Institute's president, Robert T. Pollock, declared: "The only condition imposed with this gift is that it be wisely spent for the youth of America."[25] Despite Pollock's rhetoric, this corporate sponsorship had the potential to alter his organization's science education programs. For instance, a "special joint committee," composed of American Institute and Westinghouse representatives, quickly formed, which "considered and passed on plans and provided funds regularly to the Institute."[26] In addition, the American Institute relied heavily on Westinghouse for facilities and equipment at the World's Fair: From demonstration tables and spotlights to office equipment and student lockers.[27] Westinghouse officials, meanwhile, worked to orient the science education programs to industrial and consumer applications. For example, Lufkin invited the American Institute's staff members to visit the Westinghouse plant in New Jersey so they could "visualize some of the things that the sponsor of the Junior Science Clubs is doing."[28] In January 1939, Westinghouse awarded a free trip to its Pittsburgh laboratories to Helen Miller, an aspiring scientist and prizewinner from the Science Congress, where she met with the director of its research laboratories and renowned nuclear physicist, E. U. Condon.[29]

Despite this support, the American Institute still needed the cooperation of local school leaders to supply science teachers, lab technicians, and student participants. It therefore presented a more modest request in the summer of 1938: To authorize the release of selected high school students and teachers to participate in the World's Fair. Emphasizing that the American Institute would "bear the total expense of transportation, construction, decoration, equipping, as well as rental of two sections of the World's Fair building," it hoped to secure the school board's support in "presenting to the students of New York schools a very rare opportunity for a thrilling adventure in science."[30] The organization even offered to pay for students' admission to the fair and to arrange for insurance.[31]

This plan ultimately appeased Superintendent Harold G. Campbell, who expressed his approval: "The young scientists who take part in the American Institute's Junior Science Club and who participate in the annual Junior Science Fairs and Science Congresses, by actually performing scientific research and constructing practical machines, unquestionably come to understand them better."[32] The superintendent also urged local schools to promote the event to students and offered incentives for the city's science teachers to participate. By the end of 1938, then, the American Institute's leaders had secured a corporate sponsor for funds and facilities, and the cooperation of the city school district to provide students and teachers. With these logistical arrangements in place, they began to focus more deliberately on how the science activities of youth should appear to the anticipated millions of visitors.[33]

Members of the American Institute's Junior Advisory Committee, in charge of planning for student participation at the World's Fair, viewed the coming event as an opportunity to demonstrate the effectiveness of innovative methods in science education to a national audience. At a January 1939 meeting with teachers and administrators from the local schools, H. H. Sheldon—the American Institute's managing trustee and professor of Physics at New York University—called it "the greatest educational experiment attempted." Vice President H. C. Parmelee stressed the planned science exhibit's civic benefits: "There is no force quite so great for the building of honest citizenship as the study of science and its applications in engineering." Paul Mann viewed the World's Fair as a pivotal step in pedagogical innovation, and he predicted that "in five years the Science Fairs would be the most potent means of science teaching, outside of classroom teaching, that there is in this country." Morris Meister, now principal of the Bronx High School of Science, considered it to be a critical moment for demonstrating new methods in science education to a national audience in light of the rise of fascism abroad. "We are in a state in the world history where the only solution of our problems is the heightening of science," Meister argued: "We must imbue in children the feeling that democracy is science; that intolerance is bad science; that prejudice is unscientific."[34] In these ways, the American

Institute's leaders sought to demonstrate to fair visitors that "a new method of progressive science education which has been developed in New York City in the last ten years" would strengthen democratic citizenship in the United States.[35]

Westinghouse officials articulated different social and political justifications for these science education programs. A. P. Craig, manager of the Westinghouse exhibit at the World's Fair, explained that its sponsorship of a national network of science clubs would help supply American industry with "a steady stream of well-grounded scientific workers in future years."[36] Furthermore, Westinghouse's central purpose for constructing a building at the World's Fair was to promote its electrical products and conveniences to millions of consumers. It used the occasion to launch a new advertising campaign.[37] According to the company's Vice President David Youngholm, "the exhibits will demonstrate how electricity has assumed the burden of major household tasks, and how it contributes in many ways to the pleasure, convenience, safety and health of people."[38] One such demonstration would depict two miniature models of farms—one with and the other without electricity—to dramatize "the merits of electric power to the farmer."[39] By touting these sorts of innovations, Westinghouse officials sought to highlight their longstanding rural electrification program and the company's research and development of power sources and applications.[40]

Exhibits in Westinghouse's Hall of Electrical Living, moreover, intended to impress upon visitors "how electricity has assumed the burden of major household tasks, and how it contributes in many ways to the pleasure, convenience, safety and health of people."[41] The Playground of Science would feature an array of visitor-operated objects including an infrared musical light beam, a body heat receiver, and a stroboscope. The interactive nature of these exhibits, designed to entertain visitors, would not emphasize the scientific principles at work.[42] Similar priorities informed Westinghouse representatives' plans for the Junior Science Hall. Believing that "the success of the Westinghouse Exhibit is going to depend, to a very large extent, on how this Junior Science Activity is conducted," they searched for additional methods to present science in ways that would attract— and dazzle—as many visitors as possible. Above all, spectators should appreciate the technological and consumer applications of students' projects.[43] As Roland Marchand and Michael L. Smith have argued, many industrial corporations in the interwar decades assumed that public displays of science should entertain and not intellectually challenge visitors. Similarly, as Steven Conn has shown, the corporate sponsorship of scientific expositions tended to yield uncritical praise for the wonders of technology in modern society.[44] At the New York World's Fair from 1939 to 1940, the discrepancies between science educators' civic priorities and Westinghouse officials' commercial aims would become increasingly evident.

The First Year

With its overarching theme, "Building the World of Tomorrow," the 1939–1940 World's Fair has been characterized as a deliberately educational enterprise about the value of science to society. At the same time, the event was ultimately a profit-seeking venture to revive a sluggish local economy in the last years of the Great Depression. With this primary objective, the World's Fair's organizers largely ignored the viewpoints of professional scientists. Instead, they sought to highlight the societal contributions of industrial corporations and tended to depict scientific inquiry as magical entertainment with applications for consumer products.[45]

In the summer and fall of 1939, 825 students under the supervision of 119 teachers displayed exhibits and conducted laboratory experiments in the Junior Science Hall of the Westinghouse building. These included 40 science fair projects representing the fields of physics (focusing on electronics), chemistry (emphasizing synthetic materials), biology (primarily about genetics), physiography (relief maps), nature studies, and student photographs. Two science laboratories featured students conducting experiments under the supervision of science teachers. Students also demonstrated techniques of developing pictures in a photography darkroom. Most attended New York City schools, but a few out-of-town members of the American Institute's growing national network of Science and Engineering Clubs participated as well. Largely through the efforts of Westinghouse officials, moreover, the Junior Science Hall introduced activities to attract more visitors including a student-operated amateur radio exhibit. By mid-summer, a new exhibit featuring students making ceramics drew crowds of spectators, as did a demonstration of the processes of assaying gold. Meanwhile, students from Girls Commercial High School in New York conducted chemical experiments in the manufacturing of cosmetic products behind a glass barrier in front of a crowd of onlookers. Nearly six and a half million people visited the Westinghouse building in the World's Fair's first year.[46]

Various manifestations of science also abounded beyond the Junior Science Hall. As historians Peter J. Kuznick and Robert W. Rydell have

Figure 3.1 Girls perform chemical experiments in the Westinghouse building at the New York World's Fair, 1939. Collection of the New-York Historical Society.

demonstrated, corporate visions of science in the planning and execution of the World's Fair tended to present scientific inquiry as "magical" and its outcomes as commodities for public consumption.[47] Bell Telephone, for example, presented "Voder...the machine which combines hisses and buzzes to form speech." DuPont's "Wonder World of Chemistry" simulated the manufacturing of some of its commercial products, while General Motors' "Casino of Science" escorted visitors on a "sound-chair ride depict[ing] highways and cities of 1960." In the Westinghouse building, spectators could marvel at "Elektro," a mechanical robot who performed a host of "tricks to entertain visitors."[48] Cartoon pamphlets simulated a conversation among fairgoers who marveled at the mechanical robot but concluded that "he's not nearly as wonderful as a modern electrified home." To stress this point, an auditorium featured a continuous "battle of the centuries" between two women washing dishes: One by hand and the other using a Westinghouse electric dishwasher.[49] By contrast, students' displays and activities in the Junior Science Hall seemed to place less frequent and explicit emphasis on the industrial and consumer applications of science. Hoping that they would "demonstrate to teachers the possibilities for enriched education through science thus raising the level of the general public interest in science education," American Institute officials initially appeared to prize students' active engagement in the processes of scientific inquiry.[50]

At the same time, the American Institute sought to capitalize on a unique opportunity for publicizing its own organization and reclaiming a national voice in the popularization of science. At the Junior Science Hall, for instance, thousands of visitors received booklets inviting them to establish student science clubs as part of a growing network across the United States.[51] Similarly, 20,000 special issues of *Science Observer* were on hand with articles praising the American Institute's initiatives: "The growth of science clubs must be described as phenomenal...no real project work, no creative and individual investigations in the realm of science were common until the American Institute's Science and Engineering clubs came into being." The "sudden popularity of science," as evidenced in clubs and fairs, demanded visitors' appreciation of the students' science projects on display: "Some build motors and transformers, others breed bacteria cultures for microscopic work. Still others delve into electronics, soilless [sic] gardening and plant hormones, aeronautics, insect life, metallurgy, astronomy and a score or more other phases of science." Like other exhibits on hand at the World's Fair—television, the Hall of Medicine, an oil well, and the City of Light—the students' projects were "original," emulated "the serious endeavors of matured scientists," and contributed to the nation's progress.[52]

Special events at the Junior Science Hall also recognized students' achievements. A ceremony on July 1, for example, honored Frank Pierson, a high school freshman from Flushing, New York, for effectively narrating his chemical experiments through a public address system. Lufkin praised Pierson while inviting spectators to visit the Junior Science Hall: "There are more like him in our exhibit. We would like every parent attending the Fair to see how the boys and girls of today are training themselves in the sciences for the highest type of service to society."[53] Similarly, a radio broadcast on August 12—dubbed *American Institute Day* at the fair—featured the presentation of the inaugural "Marconi Memorial Scholarship" to Robert Barkey, a recent graduate of Stuyvesant High School in New York City. During the program, American Institute President Robert Pollock highlighted his organization's membership of esteemed professional scientists: "The world of tomorrow will be made by the scientist of today lending a helping hand to the boys and girls of today," he declared, "for more than twelve years ago they started the formation

Figure 3.2 Poster promoting the American Institute Science and Engineering Clubs' exhibits at the New York World's Fair, 1939. Collection of the New-York Historical Society.

of local science and engineering clubs now known as The American Institute Science and Engineering Clubs." Pollock also made certain to credit the "great and unselfish generosity" of Westinghouse.[54] Ceremonies such as these aimed to extend the American Institute's influence in science education across the nation.

Westinghouse's publicity directors also worked with American Institute representatives over the summer of 1939 to identify new strategies for attracting larger audiences.[55] Concerned that the name "Junior Science Hall" was "too formidable and tends to draw casual visitors away," Westinghouse officials renamed it "Student Science Labs." They also explored the possibility of assigning student "barkers" to entice more visitors, searched for ways to elicit the interest of newspaper and magazine feature writers, and contemplated distributing a comic strip leaflet. They even blocked some of the outer exits to the Westinghouse building to help funnel crowds from the neighboring Halls of Power and Electrical Living.[56] Westinghouse's publicity directors from January through September released 83 news stories and arranged for 14 radio programs about their various exhibits. These aimed not only "to induce people to visit the Fair and our Exhibit," but also "to gain the attention of the stay-at-homes, so that they, too would recognize the importance of the Westinghouse participation in the Fair."[57] In these ways, both American Institute and Westinghouse representatives scrutinized the popularity of their exhibits and sought to increase their exposure.

Conflicting Priorities

As the World's Fair drew to a temporary close in the fall of 1939, American Institute and Westinghouse officials immediately began planning for the reopening in May 1940 and continued to seek modifications for enticing greater numbers of visitors. On the whole, World's Fair organizers and sponsors had been disappointed by the turnout in the initial year, and the fair itself failed to turn any sort of profit. Although nearly 26 million people attended in 1939, this number fell well below expectations, and the fair suffered from an operating deficit of $18.7 million. Furthermore, conflicts emerged between commercial interests, who wanted visitors to invest in more consumer goods, and "social theorists," who invited visitors to view American society critically. Commercial interests ultimately triumphed, as the increased pressures to make the fair more profitable in its second year, in conjunction with the looming prospect of world war, thwarted most inclinations to scrutinize the shortcomings of American democracy and industrial capitalism.[58]

Comparable disagreements about the display of science education erupted between Westinghouse and local science teachers during these intervening months. Although the American Institute depended on the cooperation of both parties, its leaders increasingly sided with their industrial sponsor because of their aspirations to promote and oversee science clubs across the nation. Through Westinghouse's financial support, moreover, the American Institute also acquired *Science Leaflet*, a weekly publication for teachers advising science clubs. New affiliations with the State Academies of the AAAS and the Junior Chamber of Commerce of the United States also reveal the American Institute's designs to accelerate its expansion.[59]

These strategies appear to have yielded significant short-term results. By November 1939, the American Institute had received nearly 5,000 inquiries from visitors to the Junior Science Hall. Furthermore, 800 new science clubs joined the American Institute in 1939, and membership more than tripled—from roughly 6,000 to 18,500 students in all states across the nation. The American Institute also continued to hold its annual science fairs at the American Museum of Natural History in 1939 and 1940. The popularity of the local fair was robust; 28,000 visitors attended in April 1940. Westinghouse's financial commitment made this expansion possible. Indeed, the American Institute's ambitions to achieve greater national prominence in the field of science education deepened its financial dependence on its corporate sponsor—a relationship that influenced how it would present students' science projects when the World's Fair reopened in 1940.[60]

Yet the immediate quest to draw more visitors to the World's Fair and the broader goal of increasing science club membership obscured the American Institute's original motive to promote innovative pedagogies in science education for civic ends. Reflecting on the relative successes and shortcomings of the recently closed fair, Hazel MacCallum, the American Institute's executive assistant for Junior Activities, underscored the popularity of demonstrations such as photography, telescope construction, and cosmetics. She therefore urged managing trustee, H. H. Sheldon, to favor displays in the coming year that would elicit the greatest public interest, such as model boats and airplanes, as well as glass blowing. American Institute officials subsequently informed local science teachers of these new criteria: "All exhibits and activities should be chosen largely because they have an entertainment value, although some may be scholarly most of them should be *easily* understood and all made very *graphic*." MacCallum specifically recommended that "'magic' from chemical experiments should be further developed" and searched for ways to "show more clearly the commercial application" of students' experiments.[61] American Institute representatives stressed this point in particular: "Activities brought to the World's Fair need to have that quality which arrest and hold the attention of the public. These are not necessarily the most scholarly exhibits."[62] This heightened attention to the entertainment features of students' projects matched Westinghouse's own emphasis on the "thrills" and "marvels" of their displays.[63]

This new stance angered local educators, however. In February 1940, Associate Superintendent Frederic Ernst complained to Sheldon that a number of science teachers, who had approved some student exhibits "for inherent science values," frequently found themselves overruled or ignored by American Institute and Westinghouse representatives. Much of this dissatisfaction stemmed from disagreements about what constituted real science: "Toward the end of the season last year it was noted that hobbies, manual art activities and plastic art activities were gradually displacing the science projects." Ernst communicated local science teachers' contentions on the matter: "If the project is supposed to be suggestive for science clubs, then the activities should be kept on the general scientific level rather than on the plane of hobbies." Acknowledging the need to capture the attention of fair visitors through dynamic displays, Ernst insisted that "it is readily possible to restrict such to scientific phenomena."[64] He also recounted some tense public confrontations between science teachers and American Institute and Westinghouse officials during the 1939 fair over the quality of student exhibits. Furthermore, Westinghouse representatives had publicly embarrassed a science teacher for using measuring devices and chemical kits from a rival company. In sum, the associate superintendent demanded that local school administrators and science teachers assume considerably greater oversight in planning for and conducting the coming fair.[65]

Sheldon promptly consulted with Lufkin to gauge Westinghouse's position. Demonstrating his disdain for some of the school board's requests, Lufkin refused to grant a committee of schoolteachers the authority to determine the criteria for selecting student exhibits. Despite the science teachers' view that "such activities as pottery-making and similar activities are non scientific," Lufkin demanded that Westinghouse officials retain the final authority in making such a determination: "We must at least have the privilege of keeping the show going." He also accused school officials of not appreciating Westinghouse's generosity. "Since the Institute are giving them this opportunity to display their wares without expense," Lufkin argued, "I think they are asking too much in wanting to take over the whole show."[66] His repeated uses of the word "show" cast the students' science exhibits and experiments as a form of entertainment. By referring to the students' projects as "wares," moreover, Lufkin essentially reduced their work to products for public consumption. Similarly, a Westinghouse press release about "these young research workers" invited the public to "share thrills of science" by pressing buttons to activate students' "animated" and "illuminated" exhibits.[67] There was no emphasis on demonstrating the civic value of students' active investigation of scientific phenomena. These conflicting priorities reflected a larger trend in the public presentation of science in the twentieth century. As historian John C. Burnham has shown, the conceptual logic and processes of scientific investigation gradually diminished in favor of its outcomes and consumer products. Such portrayals of science promised to entertain and benefit individual consumers, but the prospects for societal progress faded from view.[68]

The American Institute's subsequent response to the school board reflected its sponsor's demands. Sheldon insisted on maintaining the authority to overrule science teachers' selections of students' projects. Concerned about engaging "the lay public," he explained, "we do not feel that we should be compelled to exhibit material which has no appeal to the average spectator." As the American Institute was actively soliciting prospective exhibits from its newly affiliated science clubs across the nation, Sheldon demanded that it retain the prerogative of selecting out-of-town participants. He also prohibited the use of any laboratory equipment not furnished by Westinghouse. Sheldon urged the associate superintendent to remind his colleagues on the board of education of their mutual obligation: "The American Institute are the beneficiaries of the Westinghouse Company, without whose cooperation neither of us could achieve our aims. The interests of all three parties must be kept in mind in reaching a successful and harmonious working arrangement."[69] His refusal to accommodate local educators resulted from the American Institute's deepening financial reliance upon its industrial sponsor.

The Fair Reopens

More than four million patrons visited over 300 young scientists in the Student Science Labs at the World's Fair in 1940. Forty new exhibits were on display in glass cases, including a "capacity relay," oscilloscope, Geiger counter, iron lung, and a wooden model of a submarine escape hatch. In the laboratories, students engaged in activities that "surprise[d] Fair spectators with chemical stunts," demonstrated the manufacturing of cosmetics, depicted "soilless" gardening—and despite the objections of local educators—made pottery.[70]

To highlight its burgeoning network of science clubs, the American Institute awarded weeklong trips to science fair prizewinners from high schools across the United States. It arranged for some of these students to participate in ceremonies and to speak on radio programs.[71] After interviewing several students on NBC's *Bright Ideas Club* program of August 9, for example, the host welcomed the American Institute's new president, H. C. Parmelee, who began by articulating the purpose of science clubs in civic terms: "We are building a future generation of clear-headed, level thinking citizens." As

Parmelee continued, however, he stressed the importance of training a highly skilled labor force that would bolster the nation economically: "We are preparing them to take their places in the industries of the future."[72] As managing trustee, Sheldon had also articulated a similar rationale for improving science education in the United States: "If we are to continue to hold our position of world supremacy which we have but recently gained in science and engineering...we must recognize genius early."[73] Science clubs, according to Sheldon, would play a central role in this effort. Subsequent radio broadcasts from the Westinghouse building portrayed the Student Science Labs and the American Institute's broader network of science education programs in the service of the nation's economic and political needs. Westinghouse, meanwhile, carefully monitored the content of such publicity. In June, for instance, its promotional directors cancelled the broadcast of a student who had built a model airplane "because his talk was too technical."[74]

Students' science projects typically eschewed overt political messages, but some signs of the world war surfaced in the final days of the 1940 World's Fair. At a ceremony on September 23 for the burying of a time capsule constructed by Westinghouse, Irving Lazarowitz, a student at James Madison High School in New York City, characterized the science laboratory as a haven from human strife and a source of optimism for the future. Assigning the task of rebuilding a damaged world to his own generation, Lazarowitz praised American democratic traditions for complementing the spirit of scientific inquiry: "We know that our elders are permitted to work unmolested on their great new discoveries for saving life and making it more enjoyable because they, too, grew up under the American form of government."[75] Articles in *Science Observer* began to portray the American Institute's science education programs as pillars of the nation's political stability. Blaming self-aggrandizing "conquerors" for the impending war abroad, and targeting hypocritical "usurpers" at home for "seek[ing] to overthrow the foundations of democracy," an October 1940 article claimed that communists and fascists sought notoriety because they had performed poorly in school: "Finding it impossible to gain recognition because of excellence in their grades or brilliance in club activities." The American Institute pointed to the work of students at the Westinghouse building and its own science clubs across the nation as a productive outlet for youth and safeguard against the influence of radical ideologies.[76]

When the World's Fair shut down permanently in November 1940, the possibility of American involvement in a global war seemed far greater than it had when the fair had opened in May 1939. An article in *Science Observer* reflected this shift in linking the quality and scope of science education to the nation's political strength. Acknowledging the impending draft of nearly one million American soldiers, it asked, "in what manner can a man's hobby help in the development of a military organization?" Arguing that there is no "single hobby which could not have a military application," the article claimed that science clubs and fairs would prepare youth for productive work in the armed forces, and more generally, in all manifestations of industrial or military leadership. These priorities seemed congruent with those of the American Institute's industrial sponsor. Only several months earlier, a front-page article in *Westinghouse Magazine* had declared "National Defense becomes a part of the task for Westinghouse men and Westinghouse management."[77]

Innovative methods in science education increasingly served wartime ends as the United States moved closer to armed conflict. More broadly, mobilization began to alter popular conceptions of the duties and appropriate actions of American citizens. Rather than scrutinize and debate the strengths and shortcomings of their democratic and capitalistic traditions, Americans should prize political unity in a possible conflict against a common enemy.[78] By November 1940, American Institute officials believed that their widespread exposure through the World's Fair and burgeoning national network of science clubs and fairs had placed them in a position to lead in this effort. Despite such confident projections, they would be denied that opportunity.

Aftermath

In many respects, the American Institute realized its quest to present science to the millions of visitors to the New York World's Fair in 1939–1940. Energized by Westinghouse's financial support, it also took advantage of numerous opportunities to develop science education programs nationally. Its leaders welcomed the sponsorship of one of the nation's most powerful manufacturers and the potential opportunities to extend the reach of science fairs and other educational activities. Sheldon in particular targeted the AAAS and its affiliated Junior Academies in various states. He had met with the state representatives of Junior Academies at the AAAS meeting in Indianapolis on February 11, 1939, to offer services ranging from organizational guidance, radio broadcasts, and traveling exhibits. Although the AAAS remained hesitant, Sheldon succeeded in enlisting the participation of the Junior Academies in Pennsylvania and Iowa in the fall of 1939.[79] He also provided funds to the North Carolina Academy of Science to help organize clubs and furnish lantern slides depicting scientific work. Its secretary and treasurer, Bert Cunningham, encouraged high school science clubs in North Carolina to affiliate either with the state academy or the American Institute. Cunningham sought the American Institute's support, because he did not feel that the state academy could sufficiently galvanize extracurricular science activities in his state.[80]

Monthly issues of *Science Observer* reached club members across the United States, and American Institute officials hoped that the sharing of affiliates' activities would fortify their growing organization.[81] Coupled with the publicity gained from the World's Fair, these initiatives prompted remarkable growth. By June 1940, 700 science clubs from 44 states belonged to the burgeoning network, which would facilitate "contact with other members engaging in similar efforts, and adults who, by sharing their interests, naturally provide inspiration." American Institute officials also helped to establish community and state science centers "to make it convenient for young scientists to gather to show what they can do."[82] In February 1941, moreover, a science laboratory for local students opened in midtown Manhattan. The International Business Machines Corporation (IBM) donated the space, and Westinghouse supplied its laboratory equipment from the World's Fair. President Parmelee acknowledged the indebtedness to Westinghouse for allowing the Institute to establish "a well rounded program of junior science activities, both for its own use and as a practical example for other communities to follow." Furthermore, the American Institute's press release publicizing the opening of the student laboratory reaffirmed the civic roles its science education programs would play: "Scientists of Tomorrow Must Be Trained Today to Help Rebuild World Torn by War."[83]

In the midst of these developments, the American Institute's collaboration with Westinghouse altered the civic purposes of science education. It also proved to undermine the organization's long-term financial stability and influence. In planning for student science exhibits and laboratories at the World's Fair, American Institute officials had sometimes found themselves caught between the priorities of their industrial sponsor and local science educators. Anticipating further financial benefits, they tended to favor Westinghouse, which sometimes yielded depictions of science that bordered on entertainment and obscured their foundational quest to cultivate rational thinking for democratic citizenship. School administrators and science teachers expressed their dissatisfaction with this inclination. Seeking to entice as many visitors as possible to their building, Westinghouse officials

Figure 3.3 Shirley Gesser prepares microscopic slides at the American Institute's student laboratory in New York City, 1941. Collection of the New-York Historical Society.

promoted their industrial work in science and consumer products by entertaining visitors and emphasizing the modern conveniences and material applications of science.

The American Institute's partnership with Westinghouse also proved to be fleeting. In August 1941, Westinghouse abruptly terminated its financial support. Reeling in debt and unable to sustain the science education programs it had developed over the past decade, the American Institute was compelled to close the student science laboratory and even to cancel its longstanding annual science fairs for local youth. Furthermore, its national network of over 800 student science clubs now belonged to the new beneficiary of Westinghouse's generosity: Science Service, Inc., in Washington, DC. Science fairs, congresses, workshops, and clubs in New York City diminished, and the American Institute lost an opportunity to lead a national movement in science education.

Publishing magnate E. W. Scripps and biologist William E. Ritter had founded Science Service in 1920 as a news agency to disseminate and promote scientific knowledge, methods, and attitudes to all Americans for democracy to survive in an era of apparent irrationality and superstition.[84] Together, Westinghouse and Science Service quickly hatched a new program—the annual Science Talent Search competition—whose explicit objective was to identify, reward, and cultivate the most promising young scientists for national service in global war. This reorientation of the science extracurriculum to the technological "manpower" needs of the nation continued to utilize educational methods such as hands-on learning and student-led projects. But it marked a departure from the civic ideals espoused by the American Institute and science educators since the late 1920s. The seeds of this transformation had been planted in the Westinghouse building at the World's Fair as the United States emerged from the Great Depression and mobilized for war.[85]

According to Joseph Cusker's cultural analysis, the New York World Fair serves "as a transition point, a prism between the pre-and post-war worlds."[86] In the Westinghouse building, local science teachers and some American Institute officials aimed to elicit public support for innovative methods in science education that cultivated critical thinking and democratic citizenship. Amid conflicts over the criteria for selecting worthy projects, however, these purposes frequently yielded to the American Institute's organizational ambitions and Westinghouse's quest to attract visitors. Some aspects of students' science projects and laboratory demonstrations attempted to dazzle audiences and stressed consumer applications. Indeed, the amateur radio, photography, ceramics, and other demonstrations did present science as entertainment. Local science teachers also lost their authority to select worthy student exhibits. Furthermore, the quest to identify and cultivate science talent for military and industrial demands as the United States moved closer to war hardly resembled the American Institute's initial impetus for participating in the World's Fair. Instead of cultivating rational and empathetic citizens in a participatory democracy, these science education programs increasingly aimed to fortify national defense and a robust domestic economy.

Historians of American science education have pointed to the aftermath of World War II and the emergence of the Cold War as an era of increased emphasis on grooming high-achieving youth with intellectual capital to bolster the United States militarily and economically. According to these interpretations, "manpower," "professionalist," or "vertical" rationales for American science education in the postwar era began to rival longstanding efforts to promote a kind of widespread scientific literacy for civic and democratic ends.[87] Consideration of the science education on display at the 1939–1940 World's Fair suggests that this shift originated even earlier. The nation's mobilization for World War II and subsequent economic revival also established a long-term precedent for industrial involvement in American science education. Most notably, the annual Science Talent Search would prize the military and consumer applications of scientific research for decades to come. In the process, meritocratic purposes began to supersede democratic aims in American science education.[88]

CHAPTER 4

Enlisting Science Education for National Strength

World War II profoundly altered American schools. The nation's urgent demand for soldiers and factory workers enticed many adolescents to abandon their studies.[1] After a half century of dramatic growth, high school enrollments fell from nearly seven million students in 1940 to roughly five and a half million by 1945. Those who remained in school encountered new vocational and "pre-induction" courses, funded by hundreds of millions of federal dollars, to train those who would soon serve in the military or in war-related industries. Existing curricula—primarily in the sciences, mathematics, and physical education—frequently adjusted to meet these national imperatives as well. Some chemistry classes, for example, began to focus on explosives, gases, and plastics. Mathematics courses incorporated lessons in navigation and aviation. General science classes often highlighted meteorology, photography, and radio transmission. Vocational programs similarly introduced war-related subjects including nurse training, aeronautics, mechanics, and military preparation. Beyond coursework, more than two-thirds of the nation's high schools adopted a Victory Corps for community service and conservation of resources. Schools also served other war-related functions, including registering soldiers, disseminating ration books, selling war bonds and stamps, and assisting the Junior Red Cross. Historians disagree about the extent to which mobilization for national defense during World War II left a lasting legacy on American schools. But the war's immediate and pervasive impact is unquestioned.[2]

 Science was very much in the news during the war. As a clearinghouse for research and discovery, Science Service informed millions of Americans about the impact of scientific knowledge and innovation on the prospects for Allied victory. The organization had assumed an educational mission from its inception but did not target the nation's schools in its first two decades of existence. Watson Davis, an engineer and aspiring science writer, became Science Service's managing editor in 1923. He assumed its directorship in 1933, a position he would hold until his retirement more than three decades later. Throughout his tenure, Davis straddled a wide range of intellectual networks among professional scientists, educators, and science writers.[3] In 1937, the American Institute honored him for popularizing science through news media. Like Morris Meister, L. W. Hutchins, and H. H. Sheldon, Davis believed that democracy could flourish if citizens adopted scientific ways of thinking. In accepting his award from the American Institute, he described some of the societal benefits of widespread scientific literacy: "The great mass of people through accurate and interesting accounts of science's successes and failures can glimpse and understand that essence of science, its trying, testing and trying again." "If they build their own convictions that this is a good, sensible, successful and useful method," Davis proposed, "then there is hope that they will apply it more widely to everyday life, to our human relations, to running our businesses, to our governments, to everything that we do." Science bolstered longstanding social and political values in the United States. "The ideals we cherish, such as liberty, opportunity, the pursuit of happiness, freedom, democracy," Davis concluded, "are achieved by the utilization of scientific methods."[4] Communicating the methods of science to the public therefore assumed a critical civic function: "Test of reason and experience can weed out the charlatan, the incompetent and the unworthy in high and low places in our people's business, if we see to it that democracy is free to operate."[5]

In the mid-1930s, Davis and Science Service's executives began to express interest in the science education of American youth. Acknowledging the emerging Junior Academies of Science, they anticipated producing printed materials and recorded science lectures for what they hoped would become a national organization of high school science clubs. At the New York World's Fair in 1939, Davis became acquainted with G. Edward Pendray, a former science news editor and engineer working as a public relations executive for the Westinghouse Electric and Manufacturing Company. Davis and Pendray estimated that only 1,000 of the nation's 25,000 secondary schools had well-trained science teachers, and they lamented that few offered even basic courses in the sciences. Pendray subsequently convinced Westinghouse executives to work through Science Service as a means of improving science education for American adolescents. As discussed in the previous chapter, Westinghouse terminated its support of the American Institute, and Science Service inherited Science Clubs of America (SCA) in September 1941.[6]

Within a few months, Science Service established a new educational program: A competition for the most scientifically accomplished youth in the United States. The Westinghouse Science Talent Search became an annual event to complement Science Clubs of America's expansion. Davis and Pendray conferred in the fall of 1941 with Harvard University astronomer and Science Service Trustee, Harlow Shapley, and Science Clubs of America's director, Margaret Patterson, to determine how Westinghouse's support could reward and encourage the most worthy high school seniors. Several years earlier, Pendray had initiated a national search for ten high school seniors worthy of Westinghouse scholarships to attend the Carnegie Institute of Technology. Davis, meanwhile, took inspiration from a nationwide poetry competition for female youth. Sponsored by Camp Fire Girls, the finalists in 1940 (one of whom was Davis's daughter, Charlotte) won trips to New York City to be judged by a board of poets and English scholars. Davis believed that a comparable program should recognize the nation's most promising students in science.[7]

By March 1942, Westinghouse agreed to donate $36,000 to Science Service to coordinate Science Clubs of America and to conduct a talent search among high school seniors "which will culminate in the award of $11,600 in college scholarships to 10 boys and 10 girls."[8] Davis urged Science Service's Board of Trustees to consider the long-term benefits of furthering the science club movement—what had begun "as a rescue operation." "There is the probability," he claimed, "that it will develop into an activity of comparable importance to the 4-H Clubs. Science Clubs of America, sponsored by Science Service, provide an effective channel of cooperation between professional and amateur scientists of all ages." Davis envisioned that these educational programs would foster a broad coalition: "An activity in which newspapers, high schools, museums, state and local science academies, universities, and scientific and educational institutions, may cooperate." He pointed out that nearly 1,000 science clubs with roughly 25,000 members already belonged to Science Clubs of America. In the coming academic year, Davis predicted, "this movement should double or triple in size."[9] With Westinghouse's support and Science Service's subsequent approval, news of the inaugural Science Talent Search was sent directly to each of the nation's high schools and through the growing network of science clubs.[10]

"All Out for Defense"

Science Service's quest to launch a national movement in science education confronted widely varying curricula in high schools across the United States. Generally speaking, however, as high school enrollments ballooned in the first four decades of the twentieth century, the proportion of students taking courses in discipline-based sciences declined, especially in chemistry and physics. In the 1940s, some educators sought to reverse this trend by revising the curriculum. To make the sciences appear more interesting to students, according to these reformers, new "general science" or "fusion" courses should emphasize their practical applications to daily living. These would cultivate good citizens, who would actively evaluate the worth of various scientific innovations and products. Biology courses had already seemed to be moving in this direction; educators' recommendations for physics and chemistry would begin to take effect by the 1940s. Some professional scientists, meanwhile, did not believe that proportionally low high school student enrollments in the sciences required the same sort of curricular reform. They tended to defend discipline-based science courses as appropriately rigorous and suited only for the most intellectually capable; most students should therefore eschew the bulk of high school science offerings. According to this view, improved guidance would direct the most talented youth to science courses in helping to meet the nation's wartime demands. Two distinct purposes and curricular approaches thus stood in relative opposition as the United States mobilized for global conflict. Science education for democratic citizenship and the critical evaluation of technological products necessitated a practical curriculum that applied theoretical principles to aspects of daily living. Science education for expert leadership required rigorous, discipline-based courses with students who had been carefully selected for their academic achievements and intellectual promise.[11]

Shortly following the United States' entry into World War II, Science Service began to urge students and teachers to adjust their activities. In an address to the American Science Teachers Association on December 30, 1941, Davis asserted that science and technology would determine the war's outcome. Science education therefore became a national obligation. "Training for science is just as important in its way as a training for actual military service," Davis claimed. As a result, "especially gifted science students should...be encouraged to continue as aggressively as possible their studies, in school and out, in order that they may join as soon as possible the ranks of those engaged directly in scientific research." Because science clubs provided experiences not readily feasible in science classes, they needed to assume a central role in civilian defense. Davis proposed activities including testing city water, assisting in hospital laboratories, disseminating effective methods for fighting fires and coping with bomb explosions, issuing first aid, and operating radios. He hoped that the roughly 30,000 members in 800 science clubs across the nation would soon swell to 100,000 members in 5,000 clubs. Davis exhorted high school science teachers to lead this effort.[12]

Science Service recommended various ways that science clubs could reorient their activities. Joseph Kraus, who had helped the American Institute organize its educational programs, outlined specific tasks in issues of Science Service's weekly publication, *Science News Letter*. Kraus called on astronomy clubs to utilize thousands of homemade telescopes in scanning the skies for enemy aircraft. Students with experience using microscopes could identify civilians' blood types. Wood and metal workers could construct fireproof boxes to safeguard valuables, while those chemically inclined could search for efficient ways to extract tin from metal cans. "Regardless of the nature of your club's activities or your work as an amateur scientist," Kraus argued, "you can help yourself, your neighbors and your country remain in readiness for a surprise attack from any quarter."[13] Subsequent articles presented club members with detailed instructions on how to conserve metal by constructing electric forges, to amass hemp and sisal by splicing rope, and to administer first aid by building splints and stretchers. These activities belonged to the larger efforts to mobilize American schools for national defense.[14]

The most elaborate initiative involved enlisting science clubs in the construction of model airplanes so the US Navy could train its spotters to distinguish enemy from friendly aircraft. Navy Secretary William Franklin Knox had called on high school students across the nation to build 10,000 models each of 50 different airplanes. Army Brigadier General J. K. Cannon similarly solicited the help of science club members in supplying models for the one million men and women volunteering to work in more than 9,000 observation posts along the east coast. The US Army Air Forces I Fighter Command Aircraft Warning Service sent detailed plans to Science Clubs of America. Kraus urged affiliated clubs to participate, and he coordinated the dissemination and production of thousands of models. By the summer of 1942, Ralph T. Millet, a captain from the Headquarters First Fighter Command, acknowledged his receipt of the initial batch of model planes constructed by the students and expressed his "amazement at the excellent work which has been done."[15]

Science clubs across the United States engaged in additional war-related activities. Students at Walton Junior High School in New York City, for example, formed a "Science in Defense Club" to study methods of conservation, safety, and maintaining high civilian morale. Clubs at Ursuline High School in Youngstown, Ohio, Hoover High School in San Diego, and Harris Township High School in Boalsburg, Pennsylvania, learned how to administer first aid. Biology club members at Stuyvesant High School in New York City contemplated the wartime uses of homing pigeons. Members of the Fleetwood High School Science Club in Pennsylvania studied methods of weather forecasting and aviation.[16] "Our School has gone all out for defense," reported the science club at Reed Junior High School in Springfield, Missouri. These students participated in the model airplane program and even reorganized in a quasi-military manner: "We have a leader, captains and patrol leaders which form a council. Students have registered and have been divided into patrols. In addition, we are organizing First Aid, Fire Patrol, Messenger Service, Aviation Study, and Current Event Clubs."[17] Club members at Soldan High School in St. Louis, meanwhile, conducted experiments on microorganisms and fungi to better understand "diseases likely to become prevalent during war time."[18] Clubs at Great Neck High School in New York and Northeast Catholic High School in Philadelphia cultivated victory gardens and studied the medicinal applications of plants. Students in the science club at Jamestown Junior High School in North Dakota issued weekly radio broadcasts about the role of science in the war and publicized their rubber salvage drives.[19]

Science Service promoted these club activities as part of its larger mission to foster greater public understanding and appreciation of science.[20] Davis touted the student groups on his nationally syndicated, weekly radio program on the CBS network, *Adventures in Science*: "All SCA clubs are being directed toward helping the war effort at the same time they are learning their science."[21] He subsequently praised students for volunteering in an era of sacrifice: "Like other war-working civilians."[22] To be sure, dozens of science clubs in the United States reported activities from 1942 to 1945 without any apparent connection to civil defense.[23] Nonetheless, Science Clubs of America's sponsors recognized that the war had oriented their students to a national mission, while they fostered the public's appreciation of science and the cultivation of a younger generation of scientifically minded citizens. Indeed, mobilization appeared to contribute to science clubs' proliferation. A mere 700 clubs belonged when Science Service acquired the fledgling network from the American Institute in 1941. This number rose to 2,600 in 1943, 4,784 in 1944, and 7,419 by 1945.[24]

Science Service also aligned student clubs with federal agencies and other national organizations as a form of wartime service through its "co-projects" program. In 1944, thousands of members "gave valuable assistance in war-necessary jobs" by working with the US Fish and Wildlife Service in teaching their communities how to conserve these natural resources. Nine hundred other students completed a fire prevention course with the US Forest Service, while the Crown Cork and Seal Company coordinated the planting of 4,000 cork oak seedlings. Science Service expanded this program in 1945 to include 13 co-projects and cast it as a unique opportunity: "Training that cannot be obtained anywhere else and...the privilege of working with some of the greatest scientists in our country."[25] The US Department of Agriculture's War Food Administration, for instance, called on club members in 29 states to locate and collect milkweed as a buoyant and water-resistant substitute for kapok in military rafts and life jackets. "Java, now occupied by the enemy," it warned, "was the principal pre-war source of kapok. The immediate military requirements are 1,500,000 pounds of milkweed seedlings, entire dependence is being placed on wild strands which are widely scattered."[26] The conservation of natural resources constituted a point of emphasis for many of these cosponsored club activities. These educational programs complemented federal campaigns to encourage Americans to sacrifice at home by purchasing war bonds, rationing food, and donating blood.[27]

But it was the conservation of a different type of resource that also captured the attention of science club organizers: Scientific talent. The American Cancer Society's co-project for 1945 called on club members to collect and disseminate information about the known causes of cancer in humans. "We are short of scientists to save us in war and protect us in peace," it urged: "Research personnel needs of the future will far exceed those of the past. Young scientists must be developed."[28] Similarly, Science Service alerted club members, teachers, and the larger public to a pressing need for the development of scientifically inclined youth. The August 1, 1942, issue of *Science News Letter* proclaimed: "Shortage of Physicists a National Emergency." It conveyed the admonitions of the War Policy Committee of the American Institute of Physics that most Americans did not appreciate the importance of expertise in physics to the war effort.[29]

James P. Mitchell, director of the War Department's Civilian Personnel Division, spoke on Davis's *Adventures in Science* radio program in

1943 about high school preinduction training. Mitchell profiled new courses in electricity, radio, and mechanics that would "provide trained manpower" for the armed forces. "The job in the schools," Mitchell argued, "is to take the raw material of good American bone and sinew and brains, and convert it without a waste motion or a waste minute into the kind of men we need."[30] Scientifically proficient women were targeted as well. *Science News Letter* featured articles about the Navy's call for college-educated women to operate radios and the American Society of Mechanical Engineers' quest for more female engineers, particularly in drafting.[31] "Qualified women are failing to take advantage of the free scientific training now being offered throughout the country," Davis lamented on his radio program in 1943. "Scientifically trained women, particularly physicists," he warned, "are urgently needed to replace men leaving for the armed forces."[32]

Science club organizers across the nation similarly claimed that their initiatives would help to produce a new generation of scientists. G. W. Prescott, professor of Biology at Albion College, founded Michigan's Junior Academy of Science in 1942 to address the state's shortage of scientists, mathematicians, and skilled craftsmen.[33] The editor of the *Pittsburgh Press*, Edward T. Leech, took pride in cosponsoring the city's science fair for youth: "Now, more than ever before, we are aware of our great debt to our chemists and engineers and physicists and biologists." "It is they who give the airplanes, the bombsights, the torpedoes, the sulfa drugs, the increased food supplies, and the machine guns with which we shall win this war," Leech declared. Investment in science clubs and fairs was therefore critical: "We must not forget that these American scientists were, only a few years ago, boys and girls experimenting and tinkering in their home workshops and in their high school science laboratories."[34] Science Service similarly touted former club members' contributions to the nation's demand for scientific expertise: "In the armed services they become technicians with valuable basic essentials already learned; in war industries they are taking their places rapidly in responsible war jobs." Those with the privilege of pursuing higher education, meanwhile, "are in training for the research jobs that will keep this country in the forefront of science."[35] After-school science programs, according to these proponents, had assumed a vital role on the home front.

Prospecting for Future Scientists

World War II similarly informed the purposes of the new Science Talent Search.[36] Westinghouse's executives believed that this national competition constituted a worthy investment that complemented their educational philanthropy including college scholarships for their employers' sons and exceptional engineering students attending the Carnegie Institute of Technology. The company also issued 4-H Club scholarships to distinguished boys and girls in rural extension programs with projects demonstrating the benefits of farm electrification. During the war, Westinghouse selected 37 women among 275 applicants from across the United States to become engineering assistants "on the basis of engineering aptitude determined by specially devised tests." These female trainees enrolled in a 36-week course of study in the fundamentals of engineering at the Carnegie Institute of Technology.[37] On October 16, 1944, the company created a philanthropic arm—the Westinghouse Educational Foundation—to further its sponsorship of American science education. Citing the popularity of their existing educational programs, Westinghouse's executives believed that the foundation would yield "increased prestige" and "widespread favorable publicity." In particular, they hoped to encourage a "more favorable attitude" toward Westinghouse "on the part of college men who will ultimately become key executives of customer companies."[38] In its early years, the vast majority of funding appeared in the form of college scholarships, as Westinghouse officials believed "that technically trained young people are a vital national resource." The foundation donated a total of $106,890 in 1944; the talent search received $40,000 of that sum.[39]

Science Service and Westinghouse benefited mutually from their collaboration.[40] For example, Westinghouse placed a series of full-page advertisements in *Science News Letter* from 1942 to 1943. These invariably outlined the company's contributions to the nation's defense and domestic prosperity by linking its consumer products to war materials. One advertisement reminded readers that "the same skill and ingenuity that are building those turbines for the merchant fleet, not long ago built more efficient electric refrigerators and washing machines."[41] Another explained that one of Westinghouse's elevator factories now produced naval guns.[42] To help readers feel more secure at home, Westinghouse's prompt and reliable circuit breakers would prevent damage to transformers if an enemy spy were to try to sabotage power lines.[43]

Westinghouse also promoted the talent search in Science Service's weekly publication. The initial announcement, titled "Wanted: Future Faradays and Curies," depicted a young man and woman, both in professional attire, gazing down at a pastoral river valley with power lines.[44] An advertisement for the second talent search sought to entice high school seniors by projecting unbounded career opportunities: "Essays and science experiments begun in high school, pay off—from here they go to college...do further research in big university laboratories...research that will build a brighter future for the world."[45] Other examples featured photographs of talent search winners visiting the nation's capital, highlighted the students' future usefulness to the nation, and touted the high percentage of award recipients who furthered their studies in college.[46] By December 1943, Westinghouse neatly outlined its corporate mission for *Science News Letter*'s readership. Scientific research in its laboratories helped to manufacture products for winning the war. Scholarships and training programs rewarded youth possessing "the native skill and talent that have made America great and will make it greater." Finally, Westinghouse's consumer products would furnish a prosperous and abundant postwar world.[47]

Westinghouse and Science Service's leaders agreed that a systematic quest for the nation's most scientifically talented youth would help the United States win the war. "We must begin an intensive search for genius, or at least, superiority in science," Watson Davis told the General Science Association of New York on February 28, 1942. "We must see to it that the unusual boy or girl gets an opportunity to go to college or technical school," he urged, "and is channeled into a definite specialized responsibility in our growing national machine for fighting and producing." The new talent search could locate and encourage the next generation of scientific leaders, who would contribute their expert knowledge and technological innovations. Such a scholarship program, according to Davis, must attempt to select the most promising youth in high school (or earlier) and without bias: "Regardless of whether their parents are rich or poor."[48] This meritocratic mechanism and objective would inform Davis's articulation of Science Service's educational programs for youth throughout the war and beyond.

The architects of this annual competition therefore devoted considerable attention to their methods of selection. Science Service enlisted Harold A. Edgerton, an educational psychologist and director of the Occupational Opportunities Service at Ohio State University, to assume the lead in devising the selection criteria. He was assisted by Steuart Henderson Britt, a lieutenant in the United States Naval Reserve. Edgerton and Britt began by defining ideal characteristics of prospective winners: Academically proficient with some scientific knowledge, socially adept, resourceful, and demonstrating initiative. "This picture of a potential scientist," they explained, "suggests a person who is intellectually capable, interested in science, and a leader among his fellows."[49] At the same time, Edgerton and Britt faced logistical limitations. With only a month to identify their criteria in advance of the inaugural talent search in 1942, they searched for methods that could be implemented uniformly and promptly in secondary schools across the nation at minimal cost that could be evaluated objectively and efficiently by Science Service's panel of judges.

Edgerton and Britt established a series of criteria to narrow the pool of entrants from the thousands of high school seniors who submitted complete applications to 40 winners and 260 honorable mentions. These included a science aptitude examination, a student's high school scholastic record, teacher recommendations, and a 1,000-word essay. This "successive hurdle" scheme meant that judges would systematically eliminate candidates by evaluating these four components of an entrant's application in this particular order. Only those scoring high enough on the aptitude examination, for example, would have their high school records reviewed by the judges. In the inaugural talent search, the examination consisted primarily of multiple-choice reading comprehension questions about various science topics. In subsequent years, it maintained a multiple-choice format, but with a greater emphasis on solving scientific and mathematical problems. Edgerton and Britt explained that they sought to identify "those who have the aptitude to study science in colleges and universities," without prizing a student's specialized knowledge in science.[50]

A student's academic record revealed class rank, the amount of science courses taken, and the number of students in his or her senior class. Recommendation forms instructed teachers to specify a student's personality traits including work habits, social skills, resourcefulness, mechanical ability, and "scientific attitude." In 1942, entrants composed an essay on "How Science Can Help Win the War." Students in subsequent years would be asked to write about scientific investigations they had conducted. The 40 winners received an all-expenses-paid, five-day trip to the nation's capital as part of the Science Talent Institute. During part of their visit, students endured interviews "aimed primarily at exploring how well the contestant was fitted for a promising career in science," and an unannounced final examination, which helped the judges determine the top scholarship winners.[51]

Davis made every effort to publicize this competition and to articulate its importance for the nation's welfare. On July 4, 1942, he opened his *Adventures in Science* radio program with the following remarks: "In accord with the spirit of the day, we are going to hear some ideas on how science can help win the war. Those who are going to make these suggestions are a few of the boys and girls just out of high school who have won Washington trips in Science Service's Science Talent Search." Davis invited six of the finalists to speak on the 15-minute broadcast, and he urged his audience not to discount the scientific promise of these high school students: "Remember that Perkin was in school and age 18 when he discovered the dye, mauve, and Newton was 19 when he worked out the principles of gravitation. So we should not be surprised if these boys and girls...have some interesting and even important ideas."[52]

Davis's guests spoke for one to two minutes each about how science could help secure Allied victory. Their suggestions included manipulating isotopes of uranium to employing infrared sensors to detect enemy warplanes. Beatrice Meirowitz of Walton High School in New York City, for example, described the possible uses of Uranium 235, "a tool of both wonderful and frightful possibilities." Nathaniel Halberstadt from Sewanhaka High School in Floral Park, New York, recommended antisubmarine and antimine devices to safeguard supply lines. While acknowledging the possible horrors of chemical warfare, Lester Hollander of Bronx High School of Science in New York City felt "thankful that our country is among the leaders in the ability to produce and invent the various gases, incendiaries and munitions necessary to the winning of the war." Davis praised the students' applications of science to military uses and noted that "such suggestions will be relayed to the National Inventors Council for their consideration." He closed the broadcast by proposing that these high-achieving high school students could someday safeguard the nation as scientific leaders.[53]

Davis consistently argued that the search for scientific talent was critical for the nation's welfare and that it complemented Science Service's larger mission. Given the war's high stakes and because "real ability for creative research and engineering is rare," the talent search represented "more than a scholarship contest."[54] Mobilizing the home front required American high schools to place an unprecedented emphasis on science courses: "Almost every boy and many girls who are at all capable of mastering these subjects are taking them in order to contribute their maximum to making America strong." As a result, Davis concluded, "we must search for, discover and nurture scientific talent among our growing generation that must contribute in a major way to winning the war and making peace safe for our civilization."[55] At the same time, Science Service would continue its longstanding mission to foster greater public awareness and appreciation of the value of science to society. Davis reported to Science

Service's trustees in October 1942 that nearly all of their educational activities—Science Clubs of America, the talent search, providing low-cost newspapers and texts for high school preinduction courses, and collaborating with the National Inventors Council—aimed to further the war effort.[56]

The theme of the talent search in 1943 was "Science's Next Great Step Ahead," and students' essays addressed the more general question of what science could do for the future. In rewarding exceptional youth, the talent search aimed to heighten Americans' awareness of science's role in the war and "to focus the attention of large numbers of scientifically gifted youth on the need for perfecting scientific and research skill and knowledge so that they can increase their capacity for contributing to the task of winning the war and the peace to follow."[57] Participating in the search, according to Science Service, approximated a patriotic duty: "In times like these any boy or girl with scientific ability, who does not utilize it to the fullest, fails to serve to the fullest extent the nation and the world."[58] Davis again featured some of the student winners on his radio program, and he profiled their subsequent work experiences in *Science News Letter*. Davis reported that many of the winners from the first two talent searches were already "making real contributions to the winning of the war through research in laboratories, in war plants and in the armed services." One had earned a rare draft deferment to conduct research in a physical chemistry laboratory at Cornell University; another spent his summer assisting in experiments on radium. Marina Prajmovsky, one of the top scholarship winners from 1942, spent her summer vacation after her freshman year at Radcliffe College as a chemical analyst at the Naval Research Laboratory in the nation's capital, where she ascertained the chemical composition of samples of "war materials."[59]

At the Science Talent Institute each spring, the 40 winners toured laboratories, dined with Congressmen, visited the White House, and interacted with some of the nation's preeminent scientists. These distinguished high school students were frequently urged to direct their interests to the United States' military needs. Leonard Carmichael, president of Tufts College and director of the National Roster of Scientific and Specialized Personnel, informed the 40 winners in 1943 that they had a "patriotic duty to advance as rapidly as possible in scientific proficiency...to gain in professional knowledge in science and engineering and thus be able to serve the nation through your specialized skills."[60] J. W. Barker, dean of Engineering at Columbia University and special assistant to the secretary of the Navy, similarly pointed to new scientific opportunities resulting from the war: "The Navy is vitally interested in developing all those who possess scientific and engineering aptitudes to the very highest possible extent."[61] M. L. Wilson, from the Office of Defense Health & Welfare Services, encouraged the students to appreciate the strategic role of food in military conflict: "It is a weapon, just as guns and ammunition, for men cannot fight when they lack strength to march."[62] In 1944, Karl T. Compton, president of the Massachusetts Institute of Technology, warned of a looming shortage of highly skilled scientists and engineers—a critical liability in the nation's security. "Battleships, aircraft and artillery of today will become obsolete," Compton argued, "but a nation possessing a great reserve of scientists and engineers can mobilize them and create still more powerful weapons of defense and offense as the need arises." "Such a great reservoir of science talent" could comprise an incredibly powerful army.[63]

With the 40 student winners in attendance, Davis broadcast his *Adventures in Science* radio program from the Science Talent Institute on March 3, 1945. He conveyed that political, military, and scientific leaders appreciated these high school students as valuable resources for the United States. The students learned about legislative plans for science research projects upon the war's end, which, according to Davis, would "help protect our nation against all future aggressors."[64] His guest was Rear Admiral J. A. Furer, coordinator of Research and Development in the Navy. The admiral complained that some former talent search winners were serving in combat units and thus unable to apply their scientific acumen to the war effort. Furer also expressed his fear of a looming shortage: "Competent scientists are so few in number compared to the total population of the country that it is especially important that they be conserved for research work." As a result, national legislation should grant college scholarships to the 4,000–5,000 most promising young scientists and exempt them from military service. Such measures could "insure our position in the scientific world ten years from now." "Otherwise," Furer cautioned, "we will be faced with a very serious shortage of scientists when the time comes."[65] Systematic competitions like the Science Talent Search could determine who would be most deserving of that privileged status.

Others placed less emphasis on science's military applications. E. U. Condon, associate director of Westinghouse Research Laboratories, warned the student winners in 1944 that the wartime neglect of basic scientific research was compromising the quality of science education for the coming generation. A peaceful and prosperous postwar world,

Condon contended, required greater popular "support for and appreciation of scientific research as an element of basic culture."[66] The US Surgeon General, Thomas Parran, exhorted the students to "join us in the fight for human health and happiness" without consideration of political borders. Even in wartime, Parran emphasized the humanitarian purposes of discovery and expertise: "Science can be used with the same revolutionary effects in *saving* life as it has been used to destroy."[67] Anthropologist Margaret Mead similarly urged the female talent search winners in particular "to take responsibility for the way in which all science is applied to the solution of human problems."[68]

As the war reached its latter stages, some talent search winners proposed that scientific knowledge could contribute to a more peaceful and prosperous future. For example, Ruth Hulda Miles of Union Free High School in Fennimore, Wisconsin, indicated in her 1944 prize-winning essay that the federal government's call for more women workers in science had inspired her medical research project to save lives and not contribute to killing. Citing the promise of penicillin, Miles declared that the current generation of high school students would "delve deep into the wonders of science and find more ways of relieving the world of pain."[69] Davis echoed these sentiments in claiming that a war-torn world was in greater need of rational, creative, and empathetic scientists than ever before: "Science is a prime agency in rebuilding civilization and outmoding war just as it has been a powerful means in fighting to save our way of life and bring victory."[70] By 1945, Davis saw Allied victory as imminent and began to call for a united postwar world. Many of the essays from talent search prizewinners in that year emphasized the reconstructive capacities of scientific knowledge. "Our future is safe," Davis proclaimed, "so long as this is the attitude of youth."[71]

Even in wartime, Davis attempted to further Science Service's founding principle: Greater popular understanding and appreciation of science would cultivate rational thought and societal progress. "The methods of science will make democracy work if they find their way to the public," he argued: "Freedom to practice the scientific method in the everyday world as well as in the laboratory is of importance equal to freedom of the press and of assembly."[72] These civic ideals had informed the founding purposes of science clubs and fairs pioneered by Morris Meister in the late 1920s. As Davis's guest on *Adventures in Science* on March 28, 1942, Meister linked science teachers and clubs to a future era of postwar peace and justice when "overwhelming numbers of our people understand the part that science plays in their lives."[73] Davis articulated this message more prominently in the last months of the war. Without science clubs developing rational citizens, he warned, "we shall not be able to protect ourselves against the forces of ignorance, brutality and insanity that have tumbled into bloody ruin so much of our civilization."[74] Davis called on science and mathematics teachers to extend their influence beyond the classroom. "An intelligent citizenry can come only from years of training," he acknowledged: "But our struggle upward toward the truly democratic way of life is based on the ability of every citizen to know and apply the scientific method of thinking to daily life."[75] In other words, the war's outcome depended on scientific leaders whose creative expertise would secure a strategic advantage. A peaceful and prosperous postwar world, however, required rational citizens who recognized the value of science to their everyday lives and supported those scientific leaders in solving perennial human problems.

Establishing a Precedent

In an address to the New York Academy of Public Education in 1943, Harvard University President James Bryant Conant contended that a free and prosperous postwar United States rested on a meritocratic system of education that cultivated the most scientifically elite leaders. "A continuing flow of well-trained, talented youth," Conant declared, depended upon a "truly universal system of education which enables the gifted boy or girl to complete the long process of scientific education without regard to the accidents of geography or birth."[76] A year later, United States President Franklin D. Roosevelt expressed a similar conviction when commissioning Vannevar Bush, director of the Office of Science Research and Development, to identify wartime precedents for coordinating scientific experts' contributions to the nation's long-term welfare. Roosevelt instructed Bush to envision a comprehensive educational plan "for discovering and developing scientific talent in American youth so that the continuing future of scientific research in this country may be assured on a level comparable to what has been done during the war."[77] This directive yielded Bush's celebrated report from 1945, *Science—The Endless Frontier*, which called for a National Science Foundation. It also addressed the question of locating and rewarding scientifically talented youth, and those recommendations drew heavily from the Science Talent Search.[78]

Bush appointed Henry Allen Moe, secretary general of the John Simon Guggenheim Memorial Foundation, to lead a subcommittee on the discovery and development of scientific talent. This group consisted largely of university faculty and administrators; Conant, Watson Davis, and Harlow Shapley were among them.[79] Shapley, who had helped to conceive and operate the talent search, advocated the twin goals of promoting greater public appreciation of science and eliciting interest among boys and girls with the promise to become scientists. He called for a publicity campaign that dramatized science's battles against human suffering, showcased scientific research in public exhibits, and provided for fellowships and science museums. Shapley sought to entice "curious and ambitious" young minds to join in "a national quest to identify the unknowns that science needed to resolve."[80] The astronomer also advocated for a national system of science awards and pointed to the example of the Science Talent Search, which "has been able to awaken a nation-wide interest among young science students."[81] Further development of Science Clubs of America in the nation's high schools, he argued, would complement this aim as well. In many respects, Shapley's points of emphasis mirrored Science Service's array of popularization and education programs.[82]

Davis recognized that much was at stake for his own organization, and he directed Moe's attention to Science Service's educational initiatives. Noting that roughly 7,000 of the nation's 30,000 secondary schools had affiliated science clubs, he proposed a national program of extracurricular activities as the most expedient method of reform. Campaigns to revise the school curriculum faced too many obstacles, and students could "best receive the inspiration and experience of scientific inquiry outside the classroom although principally within the school."[83] Furthermore, the various junior science academies, institutes, and museums already aligned with the science club movement could organize "state or regional science fairs and projects that would serve as a stimulus to science-minded youth throughout the junior and senior high school years and give orientation to the work of the science clubs of their region."[84] An expanded program would require substantial funds "to allow more extensive service to the clubs in the form of free materials and literature, the assistance of traveling club specialists, and the national organization of projects, exhibitions and field trips."[85] Davis clearly hoped that Science Service would assume the lead role.

It was therefore encouraging that Conant exhorted Moe to consider the Science Talent Search as a fruitful precedent for a national plan of federally funded college scholarships. Conant had long worried that American schools did not always reward the most deserving students. He complained that there was "little or no correlation in inherent ability as measured by performance in school and economic and geographical status which determines the composition of our colleges and universities."[86] Harvard's president also lamented the prevalence of high schools that failed to engage and retain scientifically talented students. By contrast, the Science Talent Search, Conant's National Scholarship Program at Harvard, and recently implemented military examinations for new soldiers provided valuable models for how to select future leaders—in science and other areas critical for the national welfare.[87] Moe's subcommittee would devote considerable attention to the issue of selection in the hopes of establishing meritocratic criteria and methods, and the talent search informed many of its proposals.[88]

Yet the final report eschewed any direct reference to Science Service or its educational programs. Moe deleted an appendix authored by Shapley that had acknowledged Science Service's coordination of clubs and the talent search. Shapley subsequently protested to Moe that "the extent to which the ground-breaking work of the Westinghouse Science Talent Search is ignored amounts almost to unethical procedure." He communicated Davis's displeasure as well. Shapley found the omission especially egregious because "in practice we seem to be planning to follow many of the procedures adopted by the Science Talent Search in Washington during the past four years." Not mentioning the more than 7,000 affiliated high school science clubs made little sense to Shapley if the aim was to garner greater support for a publicly funded national program of science scholarships. "The youth we seek," Shapley concluded, "are now largely developing in and because of the Science Clubs."[89]

In the wake of President Roosevelt's death in April 1945, Bush's group completed its work and submitted the final report to President Harry Truman three months later.[90] The report concluded that a systematic plan to cultivate and conserve scientific talent was both necessary and feasible: "The intelligence of the citizenry is a national resource which transcends in importance to all other natural resources." In addition to bolstering national defense, a healthy pool of scientists could ensure "good public health, full employment, and a higher standard of living after the war." A "National Science Reserve" would be on call to mobilize in emergencies. The absence of a national policy exempting scientifically talented men from military service was, therefore, alarming and contributed to a looming crisis. Informed citizens surely understood that global war and domestic problems indicated that "the current need for creative brains is not being met." The report projected a deficit of 150,000 bachelor's degrees and 16,000 doctoral degrees in science and technology fields from 1941 to 1955. "The future of our country in peace and war," it warned, "depends on that premium crop."[91]

Despite this urgent need, *Science—The Endless Frontier* claimed that most American youth lacked adequate educational opportunities. Many deserving adolescents were not receiving "the scientific engineering training which they merit and which the good of the Nation requires that they obtain." State, local, and private sources of financial support proved to be insufficient. Federally funded undergraduate scholarships and graduate fellowships were essential for establishing a meritocratic system, because "high ability, adventurous talent, is not born only into families that can pay for its development." Mitigating factors including physical health, motivation, and mental well-being admittedly clouded predictions of future scientific contributions. A program of nationwide competitions for science scholarships and fellowships nonetheless best suited the "constitutional Republic" of the United States: "There [must] be no ceilings, other than ability itself, to intellectual ambition. Every boy and girl shall know that, if he shows that he 'has what it takes,' the sky is the limit." "This is the American way," the report concluded: "A man works for what he gets."[92]

The plan for a national science reserve derived from contemporary studies about patterns of attrition along an educational pyramid. Many of these demonstrated that students from lower socioeconomic backgrounds enjoyed fewer educational opportunities. In the quest to locate and support the few who possessed exceptional creative capacity in science through graduate study, however, Moe's proposal did not attempt to account for a student's low social or economic standing. Its selection criteria included a student's score on a test of scientific promise, high school academic record (especially class rank), an inventory of activities and interests, and recommendations from teachers and principals about the candidate's ability and personality. These guidelines closely resembled the evaluation rubric for the Science Talent Search that Harold Edgerton and Steuart Henderson Britt had created three years earlier.[93]

Unlike the architects of the talent search, however, Moe's group acknowledged the salience of a student's home and school background: "No aptitude tests are 'pure' and uninfluenced by previous training. Consequently individuals attending 'good' schools are likely to be somewhat overrated by their test scores." The significance of students' class ranks similarly depended upon the quality of their high schools. Therefore, selection committees in each state consisting of three scientists, one faculty expert in student guidance, and one representative from secondary schools should weigh these societal factors in allocating undergraduate scholarships. In these ways, Moe's recommendations tried to reconcile an awareness of unequal educational opportunities with seemingly objective measures of achievement.[94]

Although *Science—The Endless Frontier* did not acknowledge the Science Talent Search, Davis endorsed Bush's report upon its publication in 1945. He featured a lengthy article by Bush in *Science News Letter* outlining the report's call for a national science foundation.[95] Davis subsequently echoed Bush's warning that "THE greatest and most critical shortage in America, when viewed from a few years in the future, is the lack and wastage of scientific talent."[96] He lamented the enlistment of talented future scientists into the military and urged national legislation to exempt such candidates from combat. Moe's plan for training 6,000 future scientists each year, Davis argued, was well worth the estimated annual cost of 29 million dollars: "A sort of insurance premium for the nation against stagnation in invention, scientific discovery, and industry, and an investment in national defense."[97]

Davis characterized the Science Talent Search as a smaller-scale precedent and touted its selection techniques. He noted that many of the winners from the inaugural contest were already serving the United States as scientists—some even in military projects. Scientific ability thus resided in all corners of the nation: "In the big cities, the small towns and the farms, in those whose parents are poor and in those with millionaire fathers or mothers, in those born here and those who came to our land as refugees." Science Clubs of America, with its 150,000 members in more than 7,000 secondary schools, would serve as breeding grounds for the nation's scientific leaders. "How good a job they will be able to do in building us all a better future," Davis concluded, "will depend in large measure on how thoroughly America searches for latent science talent and whether this search is supported with the necessary dollars and intelligent planning."[98]

With Allied victory in August 1945, the war's pervasive impact on Americans living at home had become evident. Although the United States had largely been spared civilian casualties, many facets of domestic life had changed. In addition to the more than 15 million Americans serving in the military, another 15 million people had left their homes for other parts of the nation to work in war-related industries. These unprecedented demands for weapons and supplies had spurred massive economic growth and sharply reduced the nation's unemployment rate, which, in turn, quieted many Depression-era critics of industrial capitalism. Like many other aspects of American society in the early 1940s, science clubs, fairs, and talent searches also had mobilized for national defense.[99]

To what extent, then, had the war fundamentally transformed the purposes and activities of public schools in the United States? Many of the federally funded preinduction programs ceased in the months following the war's end, as did schools' efforts to coordinate various conservation and other service activities. "As the nation demobilized," historian Charles Dorn has argued, "so did schools' extracurricular activities." Furthermore, even when schools had mobilized for the war and stressed students' patriotic duty to the nation, they never abandoned a longstanding quest to foster active and critically thinking citizens in a democracy.[100] By contrast, others have suggested that "scholastic nationalism," with the goal of molding uniformly patriotic and dutiful citizens, displaced the schools' democratic mission and set an enduring precedent for decades to come. Public education for national security, in other words, persisted as the dominant political rationale.[101]

In the realm of science clubs, fairs, and talent searches, World War II did introduce a lasting concern about shortages in scientific expertise. Science Service's school programs in the years following the war would continue to enlist talented youth for the nation's military and economic strength. In particular, they would also place greater emphasis on locating expert minds who would navigate the United States through a new global conflict in an atomic age: An ideological and technological battle against the Soviet Union. The quest to cultivate a rational citizenry for a healthy democracy did not disappear in the postwar era. But it would remain secondary.

CHAPTER 5

Sustaining Mobilization in an Atomic Age

Shortly following the United States' atomic bombings of Hiroshima and Nagasaki, and Japan's subsequent surrender in August 1945, Science Service's staff writer, Frank Thone, contemplated the future of international relations. Gunpowder had once ended centuries of feudalism, Thone argued, and the United States' harnessing of atomic power resulted from enormous financial investments and remarkable intellectual capital. He therefore predicted that only the wealthiest nations could realize atomic technology: "If cannon were the final argument of kings, atomic power is the last word of great powers." Aside from the United States and Great Britain, the Soviet Union and China could ultimately develop atomic weapons, which would become catastrophic if political relations deteriorated. For this reason alone, Thone urged, "it would seem the better part of sanity, to look and hope for a turning of all powers, great as well as small, along the road of peace made possible at last by an abundance of power for all."[1] Watson Davis echoed Thone's admonition several months later: "How successfully this situation is handled from an international standpoint will largely determine whether the world will have another war in 10 to 25 years."[2]

Leading voices in American science education articulated a host of civic justifications for developing science clubs, fairs, and talent searches in the new atomic age. The longstanding quest to cultivate rational citizens persisted. Science clubs and fairs could yield a healthy democratic society at home, and perhaps foster global harmony. More broadly, people could appreciate the rapidly increasing value of science to society in a postwar world with atomic technology. Immediately following the war, moreover, many atomic scientists believed that the nation's monopoly of atomic knowledge would be temporary and campaigned for an international body to control its uses.[3]

Deteriorating relations between the United States and the Soviet Union in the latter half of the 1940s, however, rendered it increasingly difficult for many Americans to adhere to such internationalist visions. Soviet incursions into Eastern Europe and evidence of espionage, coupled with the United States' diplomatic strategy of containment, military escalation, and furtherance of atomic tests, contributed to the emergence of the Cold War. In 1949, the Soviet Union's detonation of its own atomic bomb and the communist revolution in China fueled many Americans' apprehensions of a global communist conspiracy and appeared to justify developing more atomic weapons. The Korean War intensified these fears a year later.[4] In this political context, many science educators argued that extracurricular programs for American youth would help secure the nation from its enemies. Talent searches at the national and state levels, coupled with a new National Science Fair, would cull the elite scientific minds to maintain technological supremacy. As a result, the meritocratic rationales for science education frequently overshadowed democratic purposes from the late 1940s through the 1950s. It remained to be seen, however, whether effective methods were in place that facilitated equality of opportunity in identifying, rewarding, and grooming the nation's future scientific leaders.

Heightened Stakes

For some, the atomic age necessitated a national agenda in science education—whether for developing citizens' scientific literacy or finding the next generation's scientific experts. According to Watson Davis, science clubs provided precious opportunities for students to design and conduct experiments. By understanding the methods of investigation, they would appreciate the value of science to society and thereby support public research programs.[5] Yet Americans' apparent ignorance of science and technology was potentially destructive: "Hatred, like the neutrons from fissionable material, can cause emotional chain reactions of great violence."[6] Davis lamented "the wall of military secrecy" shielding the United States' new missile research and biological warfare programs from public view. Science Service's director, therefore, called for civilian control of the Atomic Energy Commission and for the nascent United Nations to regulate the atomic bomb. Public understanding and control of the ends of scientific research would otherwise become impossible.[7]

Harlow Shapley articulated similar sentiments. Shortly following Allied victory in Europe, the Harvard astronomer accompanied a delegation of American scientists to the Soviet Union.[8] Speaking from Moscow on Davis' *Adventures in Science* radio program, he invited listeners to "minimize the differences that arise from differing social systems and from differences in historical development."[9] Davis hosted Shapley again in 1946 to promote peaceful international relations through scientific collaborations. Political borders were of no significance to scientists. "The facts and laws of science can't recognize the boundaries that men draw on maps," Shapley declared: "Whenever the military men and the political powers that be don't interfere, scientists can get together despite barriers of language, space and different governments."[10] Evidence of growing tensions between the United States and the Soviet Union prompted Shapley to call on scientists to foster better global understanding through psychological research, by eliminating starvation and inequality, and by harnessing alternative energy sources.[11] He cautioned against the United States' continued production of weapons, as vested interests could prevent "a later demilitarization of American business and government."[12] Shapley consistently lobbied for widespread scientific understanding and political leadership transcending nationalistic allegiances. This stance placed him among a shrinking group of American scientists, as the coming wave of federal anticommunist investigations muted many voices dissenting from the Cold War consensus.[13]

Others affiliated with Science Service, meanwhile, enlisted the rhetoric of national mobilization to justify reforms in science education. This tendency was especially evident in their advocacy for military deferments for scientifically talented youth and for the creation of a national science foundation: Recommendations that echoed Vannevar Bush's report, *Science—The Endless Frontier*. Thone and Science Clubs of America (SCA)'s director, Margaret Patterson, complained shortly following the end of the war that "in our eagerness to swell numbers in the armed forces we have been stripping down our own laboratories and universities and even high schools of exactly the type of intellects and skills that has made our hard-won victory possible at all." "These are dismaying deficits," Patterson and Thone declared, "for the most powerful nation in a science-ruled world to face."[14] They expressed particular concern for the male winners of the talent search, 59 of whom had been drafted for military combat since 1942. Although all had survived the conflict, their educational progress was disrupted.[15] The federal government, therefore, needed to exempt scientifically talented men from combat to contribute to the United States' technological strength.[16]

Davis similarly hoped that a national science foundation could conserve scientific talent. He arranged for the 40 winners of the fifth annual Science Talent Search to testify before the Senate's Subcommittee of the Committee on Military Affairs on March 5, 1946. Senator Harley Kilgore of West Virginia, sponsor of Senate Bill 1850 for creating a national science foundation, chaired the session.[17] Welcoming the students as "our greatest scientific asset," Kilgore argued that the systematic cultivation of a pool of high-achieving students could eventually compensate for the United States' wartime neglect. The senator pointed to the distinguished high school students on hand as evidence of a potential remedy: "They are exactly the sort of talented youths our country needs to assist and encourage...The objectives of the Science Talent Search match precisely the basic objectives of the National Science Foundation bill which we have under consideration." Kilgore predicted that the foundation would help fund competitions like the talent search and assure "the Nation of the best use of its greatest scientific resource—that is, its young scientists."[18]

Twelve of the talent search finalists read a statement in support of Senator Kilgore's proposed legislation. They described a public foundation to further their studies and "build up the ranks of our scientific 'army'." The systematic publicity of scientific discoveries would inspire practical applications and new consumer products. According to Abraham Schweid, a senior at the Bronx High School of Science (BHSS), a robust domestic economy would result: "If they [American consumers] knew that washing machines were better for them, that there were new types of bathtubs, new types of stoves, wouldn't they want to buy those things?" Students also urged the senator to prevent security measures from restricting atomic research. Jack Durell, from BHSS, reported that most of the talent search finalists hoped for a strong international body that would share all scientific information including atomic technology. Harold Zirin, from Bassick High School in Bridgeport, Connecticut, warned that secrecy would hinder both scientific progress and the prospects for a lasting peace: "Holding back this information on the atomic bomb will lead to war." He also pointed to ongoing atomic manufacturing at Oak Ridge, Tennessee, as an ominous sign that the US government sought to monopolize atomic power. In these ways, some of the talent search winners envisioned a national science foundation for facilitating widespread understanding of the social dimensions of science and the scientific study of human behavior to prevent future military conflicts.[19]

Figure 5.1 Science Talent Search winners accompany Senator Harley Kilgore on the United States Capitol subway system, 1946. Permission granted by Society for Science & the Public.

Davis argued that a centrally governed organization would furnish new educational opportunities for intellectually gifted youth. He criticized the indiscriminate enlistment of scientifically talented men during World War II as the "bull-headed pseudo-democracy" of "technically incompetent local draft boards...[which]...has robbed us of thousands upon thousands of scientists and engineers of the future whom we sorely need at the present time." All Americans deserved legal equality, according to Davis, but it was "not undemocratic to point out that there are vast individual differences...the ability to do creative scientific research is rare indeed, although it may be more widespread than many have thought possible."[20] He did not delineate the precise skills and dispositions that enabled a scientist to strengthen the nation. Davis nonetheless assumed that only a select segment of the population possessed such qualities, and that educators and the federal government must facilitate their development. In turn, those well-trained scientific minds would apply their abilities toward the military and economic strength of the United States—thereby benefiting all Americans. Science Service's director suggested that the cultivation of a meritocracy complemented a democratic society.

Other prominent science educators articulated similar views. In 1945, Philip G. Johnson, president of the National Science Teachers Association (NSTA), pointed to wartime innovations of radar, atomic bombs, and penicillin as clear evidence that science education would determine the nation's prosperity and security. Referring to the suspension of non-applied scientific research during the war as "a tragic delay," Johnson feared that the United States lacked "those most essential commodities: trained minds and the frontier knowledge necessary for the meeting of future emergencies."[21] The NSTA's 1945 yearbook, *Science Instruction for National Security*, featured a cover illustration of five Science Talent Search finalists at work on various projects "as they prepared to play their part in national security."[22] In that volume, M. H. Trytten, director of the National Research Council's Office of Scientific Personnel, argued that the United States' postwar economic health and military superiority depended on the federal support of science education. "For the good of the nation the ablest men should be trained," Trytten concluded: "The fact that each individual so trained is thereby better off personally is secondary." Scientifically talented Americans thus had a distinct duty to serve their country.[23]

As the NSTA's president from 1946 to 1948, Morris Meister supported the pending national science foundation in the hopes that it would provide more educational opportunities for deserving youth. He also urged teachers to allow the nation's future scientific leaders sufficient time for laboratory work: "Never before in our history have we needed a more intensive search for science talent and for a program of science instruction especially designed for this segment of our youth."[24] The veteran pioneer of science clubs, fairs, and congresses lamented that most schools furnished few occasions for students to conduct their own scientific experiments—programs that revealed high school students' special aptitudes. As the founder and principal of a highly selective magnet school (BHSS), Meister complained that the most scientifically promising students typically floundered in comprehensive high schools burdened by too great a range in learning abilities. Homogeneous grouping in a separate school furnished students with the necessary motivation to excel academically.[25]

At the same time, Meister proposed that universally accessible science education could empower democratic citizens in a technologically complex era. Students must comprehend the logic and processes of scientific inquiry and not merely the products of technological innovations: "The citizen who never goes beyond the stage of regarding science as a source of gadgets, magic and miracles, is also a potential tool for a dictator."[26] Meister continued to champion experimental methods for teaching youth how to approach and solve societal problems—skills and dispositions that would inform their civic roles in a volatile atomic age. "Only in and through the laboratory," he warned, "can we develop in individuals the desirable attitude which tends to base belief and conviction upon an evidence-gathering process."[27] Didactic methods of instruction, in other words, would not suffice.[28] "Much blood and substance were recently spent in an effort to protect the democratic faith," Meister mused: "The sacrifice would be tragic and useless if the scientific spirit is now abandoned, because science and democracy are related in a symbiosis."[29] Despite these recommendations, the curricular sorting of students by ability in a growing number of American high schools rendered it unlikely that most adolescents would benefit from such experiences.

Collectively, these views conveyed that science education for all students and at all levels of schooling was necessary for ensuring the rights of citizens, solving problems of mutual interest, and avoiding the mistakes that had led to global destruction. Those who touted the empowering and mutually beneficial civic consequences of universal science education in a democracy simultaneously articulated a meritocratic vision for the cultivation of expert knowledge and elite leadership. Science Service's stewardship of clubs, talent searches, and fairs in the late 1940s and 1950s embodied these dual purposes.

"Our Most Precious Resource"

In the summer of 1946, the Bloom Radio Club of Chicago became the ten thousandth group to join SCA. By the end of the year, Watson Davis counted 12,000 clubs with a membership of approximately one-third of one million boys and girls in junior and senior high schools. In 1941, there had been roughly one affiliated science club for every 43 secondary schools in the nation. By 1946, this ratio had swelled to one in three. New York State boasted the largest number of science clubs with 1,124, followed by Pennsylvania with 784, and Illinois with 551. A few others formed in the US territories as well. A total of 162 clubs operated in other nations spanning the globe in 1946; by 1949, that number nearly reached 600.[30]

Davis viewed science clubs "as an intensely local phenomenon," whose individual members determined their specific programs. At the same time, he characterized their widespread proliferation as "a perpetual youth movement, constantly renewed by the innate and undulled curiosity and exploratory spirit of those who are discovering, through doing, the world about them."[31] Science Service officials also continued to think of clubs as service organizations—to their local communities and national institutions such as the American Cancer Society and the National Weather Bureau.[32] Harlow Shapley, moreover, hoped that more adults would embrace science clubs as a productive leisure activity. The Harvard astronomer envisioned a mechanism for cultural enlightenment that taught citizens "the advantage of looking at their subjects in a broad and penetrating and cooperative way."[33]

Davis assumed that the vast majority of SCA members possessed neither the aspiration nor the ability to become professional scientists. At the very least, however, club activities would foster students' appreciation of "the experimental method, the historical background of scientific development, the content of science, the application of science to problems of community, industry, and human relations, and…the usefulness of science to civilization."[34] All would benefit regardless of their future vocation. "The merchants, mechanics, housewives, lawyers, preachers, and all others who make up our population," Davis wrote in 1946, "will get the necessary brief taste of the content and method of science that will make them more effective and intelligent citizens."[35] Those students not aspiring to scientific careers would also recognize "the requirements of science that will make them much more intelligent voters when they are faced with the problems that must be solved by all the people."[36] According to Davis, American youth were thus distinguishing themselves in a vital way: "The awe and alarm with which science is viewed by some of their elders is foreign to these well-informed youngsters."[37] Science clubs, in other words, empowered new generations of citizens. In this respect, Davis, Shapley, and Meister echoed other leading American scientists in the postwar era, who, as historian John L. Rudolph has demonstrated, believed that science education could create a more rational society.[38]

Claims that clubs could thwart international atomic rivalries prompted a movement in the late 1940s to establish an international network through the United Nations Educational, Scientific, and Cultural Organization (UNESCO). According to Shapley, traveling panels of scientists to educational institutions in developing countries would demonstrate "not only the unity and good-will message of the sciences, but also a living exciting reminder of the oneness of all mankind."[39] Science Service's educational initiatives led UNESCO officials to conceive of Science Clubs International at their Paris meeting in July 1949, where Davis represented the United States.⁴⁰ UNESCO Director General, Jaime Torres Bodet, predicted that the linking of science clubs across political borders would have far-reaching effects: "They will together have fought against ignorance and prejudice, worked methodically with ever open minds...and with their deeper knowledge of the world about them, will better understand the bonds which united mankind in a common destiny."⁴¹ Delegates from Czechoslovakia, Denmark, England, Finland, France, Holland, and Uruguay reported the creation of preliminary science club programs. Traveling science club exhibits to less developed nations were proposed, with an eye on portions of Latin America in particular. Nonetheless, Science Service's involvement with UNESCO remained brief, and the results appeared fleeting.⁴²

Indeed, Science Service's rhetoric of national mobilization frequently overshadowed its democratic and internationalist sentiments. "We have bled our country of scientists," Margaret Patterson lamented in October 1945. The United States must, therefore, "conserve science talent" as "other precious commodities."⁴³ Davis similarly sought to align Science Service's educational programs with emerging national priorities—military defense and a robust domestic economy. For example, he exhorted the graduating seniors at BHSS in 1946 to appreciate the nation's most critical asset: "Our future is more dependent on the hoarding of our talented young scientists than the accumulation and burying of vast quantities of gold."⁴⁴ Science Service's director frequently reiterated this claim in the late 1940s. "My vote for our most precious resource goes to the science talent among our youth," he told members of the Virginia Education Association in 1946.⁴⁵ The Science Talent Search would locate those with "an investigative turn of mind, inquiring attitude, love of fact and the rare gift of scientific research ability."⁴⁶ Science clubs, too, would "determine to a large degree how well the urgent national need for scientists will be answered in the future."⁴⁷ By encouraging their members to participate in fairs and congresses, moreover, clubs provided vocational guidance: "It is on these competitive levels that the student often sees for himself that he is not sufficiently competent in science to undertake it as a career." At the same time, science competitions instilled confidence in worthy students: "A modest or shy boy or girl has often come to the realization and appreciation of his own talents by these opportunities to compare accomplishments with others."⁴⁸ A carefully coordinated network of science clubs could thus facilitate a more meritocratic system of selection in the national quest for scientific experts.

These purposes intensified by 1949 with the cementing of a Cold War political consensus in the United States. Anticommunist sentiment led most Americans to believe that an atomic arms race with the Soviet Union was very likely. The federal government and public school educators began to reinforce this message to youth through newly developed civil defense programs. American scientists, meanwhile, faced interrogations about their national loyalty, and many retreated from public affairs to their private research activities and government contracts. As historian Jessica Wang has shown, moreover, in 1949 the Department of Defense and Atomic Energy Commission contributed 96 percent of federal funds for university-based research in the physical sciences. An unprecedented degree of federal support and oversight, in turn, encouraged many physicists to consider the national defense applications of their research and even to select military projects.[49] This political context increasingly informed Science Service's justifications for the talent search. Publicity in 1949, for instance, highlighted the scientific achievements of previous winners as evidence that this competition was doing its part in fortifying the nation. Providing scholarship opportunities for deserving students had become increasingly vital in an era of atomic weapons. "With this country thrust more and more into a position of world leadership in science," Patterson declared, "this successful experiment is of even greater value to the strength and security of the United States than when it was originally planned." For SCA's director, the Cold War posed a more urgent crisis than World War II.[50]

Science Service's leaders regularly enlisted national security rhetoric as they searched for additional sources to fund their educational programs.[51] A $10,000 grant obtained from the National Science Foundation in 1952 aimed to extend their organization's influence. One product that emerged was an extensive pamphlet that classified and indexed thousands of science projects. Science Service provided a free copy to every affiliated club. As the "Largest Scientific Organization in the World," SCA produced handbooks with detailed strategies for planning activities, obtaining free or inexpensive scientific equipment from various corporations, organizing science fairs, and encouraging participation in the talent search. Other materials outlined methods for constructing scientific instruments, presenting science exhibits, and conducting organic chemistry experiments at home. A popular monthly package mailed to club members, *Things of Science*, consisted of experimental kits with instructions. Issues of *Science News Letter* frequently reported on club activities and illustrated their expansion across the nation. These resources appear to have succeeded in enticing more clubs to join SCA. By the 1957–1958 school year, Science Service could boast that nearly 17,000 groups belonged to its national network.[52]

It was corporate philanthropy and not the federal government that had funded the Science Talent Search since its inception in 1942. In the postwar era, Westinghouse's executives continued to expect the talent search to create favorable media publicity and elicit respect from colleges and universities that actively recruited its winners and honorable mentions. In addition, they believed, high school educators began to think of Westinghouse "as an important educational and scientific organization, as well as a commercial concern." The five-day Science Talent Institute for the 40 winners in the nation's capital also represented a prestigious annual event in which students (and Westinghouse officials) affiliated with prominent scientists, Congressmen, Supreme Court Justices, directors of research organizations, and members of the Atomic Energy Commission.[53] In addressing the winners in 1956, the associate director of Westinghouse's research laboratories touted his company's contributions to atomic research and encouraged his audience to pursue scientific careers in industry. The company's executives—from public relations managers to vice presidents—figured prominently in the program.[54]

Westinghouse's educational initiatives extended beyond the talent search. Its foundation sponsored vocational agriculture through Future Farmers of America, a summer institute for science teachers at the Massachusetts Institute of Technology, and atomic research through the University of Chicago's Institute for Nuclear Studies.[55] The company also provided free or low-cost materials to schools across the United States. The "Little Science Series" consisted of illustrated pamphlets for secondary school students. A comic book, *Fun in Science*, portrayed basic science experiments, while *How Does It Work?* highlighted Westinghouse's contributions to the development of new power sources. The company sold classroom wall charts depicting aspects of nuclear physics. It also produced more than a dozen motion pictures and slide films for schools, which had reached an estimated two million students and teachers by the early 1950s. A 15-minute dramatized radio show, *Adventures in Research*, was broadcast on more than 200 educational and commercial radio stations.[56] Indeed, Westinghouse's executives believed that the nation's schools comprised "a big market" and that "good reputation built among youngsters is bound to pay off in the long run on sales."[57] Even the "seemingly innocent comic book" could promote the economic benefits of household electricity, radio, and the mechanization of production.[58]

By the 1950s, Westinghouse's leaders claimed that their educational programs created "a powerful reservoir of educated specialists who can keep both our domestic economy and our national security vigorously strong."[59] Reflecting the Cold War political climate, they argued that "industry must meet the economic needs of a growing population and at the same time make our country a fortress." "The onslaught of communism" created unprecedented demands for "that priceless resource, carefully trained intellect." Congressional interrogations of suspected communists at home and continued diplomatic friction with the Soviet Union abroad added urgency to these educational initiatives. Private industry, therefore, should align with schools and universities "if we are to be assured of the creative manpower to continue the technical contributions which have done so much to make our country great."[60] This educational philanthropy also served the company's interests. Opposing communism and championing a "free economy" warranted a brand of industrial and consumer capitalism that created a market for its home products and appliances. The call to strengthen the United States militarily through new weapons similarly justified Westinghouse's ongoing atomic research.[61]

Amid Westinghouse's array of scholarships, fellowships, teacher training institutes, and higher education support, the Science Talent Search represented a relatively small financial commitment. Mindful of the philanthropic activities of industrial competitors such as General Electric, however, the company boasted that no other industrial corporation sponsored a comparable program. In the summer of 1957, Chairman Gwilym A. Price announced that the annual funds for the talent search would be tripled to $34,250. This raised the top prize from $2,800 to $7,500, and all of the 40 winners would receive some cash award. From 1944 to 1958, the foundation donated a total of $5,658,040 to all of its educational endeavors.[62]

With Westinghouse's continued financial support, Science Service pursued the exceptional few who could someday fortify the nation. In the nation's capital each spring, the 40 talent search winners learned of the importance of science in national and world affairs, received praise for distinguishing themselves at a young age, and heard of innumerable career opportunities—as well as societal obligations. Some speakers urged these distinguished high school students to apply their scientific talents in ways that would foster international peace and alleviate human suffering. Others emphasized their duty to secure the technological and military supremacy of the United States. By the 1950s, this message of national allegiance overshadowed internationalist and humanitarian sentiments, which reflected the political climate of anticommunism and widespread fears of an atomic conflict.

Talent search winners frequently learned of widening scientific horizons and unprecedented career avenues. In 1946, for example, Westinghouse Research Laboratories' associate director, J. A. Hutcheson, asserted that wartime innovations such as radar would yield multiple consumer applications.[63] With the 40 winners in attendance, Davis's *Adventures in Science* radio program on February 28, 1948, forecast a pervasive need for talented scientists. Nobel Laureate and chemist Wendell M. Stanley, physicist Karl Lark-Horovitz, and Harlow Shapley each addressed the open frontiers for research in their respective fields. Recent years of robust research, they concluded, prompted countless new questions for further investigation.[64] The 1948 finalists also heard from Westinghouse's Gwilym Price, who asserted that the talent search confirmed the "American Dream," because it rewarded the most deserving youth from all corners of the nation. "Ability has nothing to do with racial origin or economic circumstances," Price declared: "Opportunity for self-development is available to all."[65] In other words, the talent search exemplified a meritocratic educational system.

These distinguished high school seniors also learned that opportunity entailed societal accountability. "Congratulations and condolences," Shapley admonished the 1946 finalists, "you have undertaken a high responsibility that will be as heavy as it is joyous."[66] In 1948, Clarence Cottam, assistant director of the US Fish and Wildlife Service, emphasized that future scientific leaders must devise ways to conserve natural resources.[67] A year later, ornithologist William Vogt predicted that humans would exhaust their natural resources if they did not control population growth. "Man is tearing down his environment and breeding recklessly," Vogt warned, "getting deeper and deeper into trouble." Botanist Paul B. Sears similarly cautioned that "so far as forests, water, soil and wildlife are concerned, it is later than most of us think."[68] With their enlightened intervention, however, these select high school students could anticipate solutions to these looming problems.

The advent of atomic technology dominated much of the discussion at the Science Talent Institute in the late 1940s. Physicist E. U. Condon advised the 1946 finalists of the possible proliferation of atomic weapons—and the obligation to divert a human legacy of destruction. "Men throughout the world can live in freedom and justice, in love and goodwill," Condon proclaimed, "they can devote their full energies to constructive application of the rational thinking to call science to the arts of peace." For Condon, this meant fostering international organizations and closer working relations with scientists from other nations—especially the Soviet Union. Otherwise, he warned, "there is no hope for peace." Military secrets restricting scientists' travel and international collaborations made little sense: "The laws of nature, some seem to think, are ours exclusively, and that we can keep others from learning by locking up what we have learned in the laboratory and not telling it to our allies."[69] Condon therefore urged that atomic technology be shared across political borders and limited to peaceful applications. His remarks ultimately reached a wider audience when US Congressman Leroy Johnson of California, who had been present at the banquet, subsequently read them into the *Congressional Record* on the House floor. Ironically, Condon's address occurred on the same day that Winston Churchill issued his famous "Iron Curtain" speech, which cast the Soviet Union as a looming threat to the Western world. A year later, President Truman would seek to "contain" the Soviet Union's international influence by asking Congress for massive aid to Greece and Turkey.[70]

Even in the midst of deteriorating relations between the United States and the Soviet Union in the late 1940s, some scientific leaders continued to call for global harmony through international scientific collaboration. On an *Adventures in Science* broadcast from the 1947 Science Talent Institute, Shapley cautioned that "our science must not be nationalistic...we must develop planet-wide concepts of the function of science in society."[71] In 1949, Basil O'Connor of the American Red Cross stressed the commonality of people around the world. "There is no such thing as American science or French science or English science or Soviet science," O'Connor told the 40 student winners, "the world of science is one world...There is only one mankind." These high-achieving students must therefore transcend "the artificial barriers erected between individuals or groups of individuals because of their race, creed, color or national origin."[72]

Intensifying Cold War tensions and security restrictions for scientific research did not dissuade some scientists from urging the talent search winners to resist a national climate of fear. "This is America, free democratic America, free scientific America," Shapley declared in 1948. He exhorted the students "as citizens and scientists" to champion "freedom of inquiry and the honor of national service."[73] When J. Robert Oppenheimer addressed the finalists in 1950, the atomic physicist implicitly criticized the rise of anti-communist interrogations and state secrets restricting scientific knowledge. Defying Cold War imperatives, Oppenheimer scolded national governments

Figure 5.2 Science Talent Search winners join Science Service staff writer Frank Thone on a broadcast of *Adventures in Science* on the CBS radio network, 1946. Permission granted by the Smithsonian Institution Archives.

for inhibiting scientists' freedoms to inquire, express skepticism, and revise their ideas. Military secrets in the United States warped scientific progress and undermined democracy: "We do not believe any group of men adequate enough or wise enough to operate without scrutiny or without criticism... The only way to avoid error is to detect it, that the only way to detect it is to be free to enquire." He urged the students to think of the spirit of scientific inquiry as analogous to a free democratic society. "The wages of secrecy," Oppenheimer therefore concluded, "are corruption." These constituted bold declarations at a time when federal investigations of suspected communists were inhibiting political dissent.[74]

By contrast, talent search winners also learned of their national obligations in a politically divided world. In 1946, for instance, M. H. Trytten explained that World War II had demonstrated the power of creative scientific talent—in the development of radar—that afforded the United States a distinct strategic advantage: "Hundreds of thousands of men mean little against such power." A looming shortage led Trytten to conclude that "at no other time has the Science Talent

Search seemed so significant or symbolic."[75] When President Harry S. Truman hosted the students at the White House on March 3, 1949, he advised them to develop their intellectual abilities for national welfare: "I hope that you will go out of here with the idea of finishing the job and becoming an asset to this great United States of America."[76] These political justifications for conserving scientific talent intensified in the 1950s. According to Watson Davis, for example, the Korean War validated "the urgency of keeping our scientific resources constantly replenished so our country will be in a state of readiness to move forward in war or peace." "The greatest resource is the talent of our boys and girls," he concluded: "It must be recognized and cultivated wherever it can be found."[77] Assistant secretary of Defense for Research and Development, Donald A. Quarles, told the talent search finalists in 1955 that "we are in a race for technologic supremacy with the Communist world." "Our long term security," Quarles warned, "may well depend on the outcome of this race."[78] As the prospects for global harmony had largely evaporated by the 1950s, these high-achieving students learned that their primary duty was to the nation.

Talent search winners were frequently encouraged to consider the military applications of creative scientific inquiry. Henry DeWolf Smyth, a member of the US Atomic Energy Commission, emphasized in 1951 that non-applied research was indispensable for subsequent weapons development.[79] By the early 1950s, the US Department of Defense had assumed a significant role in steering the nation's scientific agendas. The chairman of its Research and Development Board, Walter G. Whitman, explained to the talent search winners in 1952 that federal defense contracts sponsored more than one-third of all of the research conducted at American universities and industries.[80] Highlighting the vital role of science in previous world wars, Carnegie Institution president, Caryl B. Haskins, urged his young audience in 1956 to appreciate the Cold War's high stakes: "The holocaust that all-out atomic war would bring; the constancy and the steady malignity of the Communist menace; the inherent perilous instability of a bipolar world." As the arbiters of the United States' political fate, Haskins welcomed the students as new members "of one of the outstanding 'warrior' groups in our nation."[81] Talent search winners also toured laboratories developing weapons near the nation's capital. In 1956, for example, they observed a new "hypervelocity gun" at the US Naval Ordnance Laboratory that launched models of high-speed missiles for possible refinement and manufacturing.[82]

Democratic justifications for science clubs and the talent search did not disappear in the postwar era. Yet even messages about the value

Figure 5.3 The 40 Science Talent Search winners visit US President Harry S. Truman in the Oval Office, 1950. Permission granted by Society for Science & the Public.

of science education for strengthening rational citizenship presumed that only a select group of youth possessed the ability to produce valuable knowledge or steer science's societal applications. Furthermore, the emergence of a Cold War ideology, especially prominent by the end of the 1940s, prompted powerful military metaphors for science education. A meritocratic quest to identify and reward systematically the future scientific experts of the United States assumed increasing urgency. This purpose similarly informed the expansion of other scientific organizations across the nation who engineered educational programs.

State Programs

As discussed in chapters one and two, state affiliates of the American Association for the Advancement of Science (AAAS) began to promote high school science clubs and welcomed aspiring students to their annual meetings in the 1920s and 1930s. These programs proliferated in the wake of World War II, when State Academies of Science

Figure 5.4 Atomic physicist J. Robert Oppenheimer, Westinghouse Vice President A. C. Monteith, Science Service Director Watson Davis, and astronomer Harlow Shapley congratulate two Science Talent Search winners, 1950. Permission granted by Society for Science & the Public.

resolved to attend more closely to science education. In addition to including high school science club members in their annual meetings, they advocated revisions to school science curricula, secured scholarships to institutions of higher education, and inaugurated state-level science talent searches.[83] Leaders of state academies frequently referenced the Cold War in promoting these initiatives. Representing Virginia's Academy of Science, E. C. L. Miller warned in 1946 that the indiscriminate enlistment of combat soldiers during World War II had created an acute shortage of scientists. He also claimed that most high school graduates from the past three decades knew little about science or its societal consequences—a dangerous state of affairs in the new atomic age. State academies, therefore, needed to steer youth into scientific fields by supporting clubs and organizing talent searches. "Our very survival as a nation," Miller cautioned, "may depend on how well we do this."[84]

Morris Meister, meanwhile, sought to entice more students to scientific careers by holding a Junior Scientists Assembly at AAAS meetings. In December 1946, a panel of college students, most of whom had gained distinction in the Science Talent Search, discussed scientists' societal obligations. The students also criticized their former high schools for lacking inspiring science teachers and sufficient laboratory experiences. In the following year, a new panel addressed "the Importance of Extra-Curricular Science Activities to Science-Talented Youth." A subsequent session in 1948 called for more emphasis on the methods and procedures of scientific inquiry in the curriculum. The students highlighted clubs, junior academies of science, and science fairs as invaluable outlets for motivated students to investigate a subject in greater depth.[85]

By the end of the 1940s, over half of the nation's state academies offered educational services. Organizers in Wisconsin, for instance, distributed news bulletins and recommended activities to club members at schools in various regions of the state. Both the Virginia Academy of Science and Georgia Junior Academy of Science inaugurated lecture programs for professional scientists to speak with interested science clubs at their schools. Most state academies invited junior members to attend their annual meetings, and students delivered scientific papers to their peers as part of their own meetings. The Indiana Academy of Science designated a "best girl" and "best boy" in the state for honorary membership in the AAAS. In Virginia, two high school teachers received graduate scholarships in recognition of their active club sponsorship.[86]

Science Service influenced many of these educational initiatives. Academy members at the University of Oklahoma, for instance, reprinted Watson Davis' article, "Science is a Hobby for Youth," in an early issue of their journal, *The Oklahoma Naturalist*, which they distributed to science clubs throughout the state. Academies in Virginia and Kentucky borrowed ideas from SCA's "co-projects" program to orient their own clubs to public service. Academy leaders in Wisconsin utilized Margaret Patterson's contacts in locating teachers sponsoring science clubs. Mary Creager, a teacher at Chester High School in Southern Illinois and head of the state's Junior Academy of Science, informed Patterson that she would use SCA's support materials for club activities "to sell ideas to our Academy Council members." In 1946, the newly formed South Dakota Junior Academy of Science welcomed Patterson to a statewide education conference, where she spoke about high school science clubs. In that year, SCA also began hosting representatives from all cooperating state academies at national AAAS meetings to coordinate their activities. At the Academy Conference in 1949, Watson Davis informed scientists of "the National Program for Science Talent." "The circumstances that bring an Einstein or a Curie to the height of creative fruitfulness, revolutionizing our physical and mental concepts and enriching the world's essential knowledge," he cautioned, "have been fortuitous in the extreme." Additional educational resources and encouragement were critical for locating scientifically talented youth more systematically, and he urged academy members to support and develop their own competitions.[87]

Many state academy leaders agreed with Davis that the talent search had assumed vital importance. Virginia and Tennessee began statewide competitions in 1945, and 21 junior academies held their own by 1951.[88] Science Service provided state officials with the names of students who had entered the national talent search, which defined the pool of state-level contenders. The number and types of awards varied considerably. Some state academies endorsed the applications of award-winning students to state institutions of higher education and granted cash prizes. In Illinois, the state academy secured scholarships to the University of Illinois for six students. Others' awards were less lucrative. The Louisiana Academy of Sciences, for example, gave the top winner in 1949 ten dollars, membership in the academy, and a forum for presenting research. Those finishing in second and third place received five dollars each. Some leaders of the Iowa Junior Academy of Science, meanwhile, lamented that they could not furnish substantial awards to state talent search winners.[89]

Illinois, which began organizing science clubs in the late 1910s, had developed an elaborate array of science education programs by the late 1940s. Its Junior Academy of Science published a regular news bulletin and yearbook, and it launched five broadcasts on the radio series *Science in Our Own Illinois*. The senior academy hosted a youth hobby fair in Chicago in 1947 showcasing exhibits of the state's high school seniors who had gained distinction in the national talent search. It also held a Science Field Day in which 400 members representing 20 science clubs displayed 148 projects. Roughly the same number of students attended and exhibited at the junior academy's meeting in that year. With help from local Chambers of Commerce, 20 seniors earning distinction in the state's talent search gathered with their families at the joint conference of the state and junior academies in 1949, where they met with various representatives from universities and colleges. Lyell J. Thomas, professor of zoology at the University of Illinois and chair of the judging committee, explained that the primary task was "to see that these especially talented students were placed in universities and colleges of their choice."⁹⁰

In the South, Alabama's Academy of Science established varied educational programs for youth. Created in 1933, the Junior Academy invited representatives of high school science clubs to deliver papers or present research exhibits at annual meetings. James L. Kassner, a professor of chemistry at the University of Alabama, collaborated with high school science teachers, Clustie McTyeire and Kathryn M. Boehmer, in partitioning the state's science clubs into 13 districts and issuing junior academy news bulletins to 300 White high schools. With assistance from Science Service, the Alabama academy launched a state science talent search in 1947. Kassner also secured one four-year scholarship each from five institutions of higher education in the state, including the Tuskegee Institute. For a brief time, Alabama's Chamber of Commerce provided scholarships "to lift the economic level of the South through scientific endeavor."⁹¹ Instances of racial inequality in Alabama's state science talent searches were abundant, however. African–American students typically won fewer than one-quarter of the state's awards, and they could only use their scholarships at the Tuskegee Institute. In 1949, the White winners and honorable mentions were awarded a three-day visit to the junior academy meeting and made valuable contacts by visiting with Chamber of Commerce members, attended a session of the State Legislature, and toured the First White House of the Confederacy. The three African–American finalists, meanwhile, spent two days at the Tuskegee Institute, where they were judged further and received prizes at a special ceremony.⁹²

The concentration of professional scientific organizations in the nation's capital, coupled with Science Service's proximity, fueled the growth of the Washington (DC) Junior Academy of Science. Its senior academy began in 1948 to introduce students to professional scientists and to allocate awards. In 1952, Watson Davis helped to inaugurate 85 local high school students and 40 recent graduates as members of Washington's newly created junior academy for their scholarly excellence and for gaining distinction in talent searches or fairs. The Committee on Encouragement of Science Talent organized conferences to acquaint members with teams of distinguished scientists "to aid in the selection and planning of projects to be worked on during the current year." As the American Institute of the City of New York had done in the 1930s, moreover, the Washington Academy organized Christmas Lectures and science fairs.⁹³ It also sponsored student field trips by train to museums and planetariums in New York City and Philadelphia, which attracted roughly 5,000 junior and senior high school students each year. Scientists representing the Washington Academy also spoke at local high schools, assisted with science fairs, and attempted to support science teachers' professional development through graduate study.[94]

State academies increasingly sought to reward deserving students who would someday join the ranks of professional scientists. The Virginia Academy of Science resolved to "discover, encourage and develop scientific interest and ability among the youth of the State by a Junior Academy of Science, by science clubs, by science talent searches and by other means."[95] In Iowa, the Clinton Corn Processing Company and the Dow Chemical Company donated college scholarships to winners of the state's science talent search "to produce seniors of outstanding ability to enter the field of science to fill the new vacant positions in many fields."[96] Tennessee's Academy of Science endorsed the applications of its state winners to college presidents. "If scholarship aid were not available," its leaders contended, "some of these talented youths would not be in college at all."[97] Inspired by the Westinghouse Science Talent Search, South Carolina's academy reported in 1953 that it planned to arrange for a statewide science fair. In that same year, Indiana's academy noted that the 86 winners in the history of its state science talent search had received a total of $57,000 in scholarships. In Georgia, 250 high school students exhibited their science projects in a statewide fair in 1953, while Pennsylvania's academy received a bequest to support youth presenting scientific papers at its annual meeting. The hope among these state academies was to funnel more deserving high school students into undergraduate study, and ultimately, scientific professions.[98] A grant from the National Science Foundation and the Oak Ridge Institute of Nuclear Studies funded a national meeting in 1957 "to encourage precollege science students to appreciate science training, to participate in scientific organizations, and to choose careers in science."[99]

Although they remained somewhat autonomous, state academies of science developed comparable educational aims in the late 1940s and 1950s. Affiliated professional scientists sought to familiarize interested and talented high school students with the rigors and rewards of the profession—as a form of career guidance and to cultivate more widespread appreciation of science. Popular concerns about the nation's apparent shortage of scientists also prompted many state academies to act. They benefited from educational networks already in place because of Science Service's talent search, the SCA, and by 1950, the National Science Fair.

The National Science Fair

Science fairs became increasingly popular throughout the United States in the postwar era. The American Institute, which had pioneered the movement in the late 1920s and 1930s, endured financial hardships that compelled the suspension of fairs in New York City. With the cooperation of the Board of Education of the City of New York, the American Museum of Natural History, and the Federation of Science Teachers Association of New York, science fairs resumed sporadically in the late 1940s and 1950s. Over five days in December 1946, roughly 2,000 students showcased 318 exhibits to approximately 35,000 teachers and students in Madison Square Garden. Special exhibits by the US Army and Navy highlighted the military applications of science to the nation's security.[100] Morris Meister organized the American Institute's science fair at Brooklyn Technical High School on December 6, 1952. A total of 350 local students, ranging from those attending kindergarten to senior high school, showcased their science projects to roughly 4,000 parents and teachers. In the afternoon, representatives from Westinghouse and General Motors presented their own "science shows" in the school's auditorium. Despite continually sparse resources, the American Institute's trustees determined to host science fairs to contribute "to our country's technical manpower program by stimulating the interest of our youth in science." As a result, fair organizers must have felt gratified that 1,831 students exhibited at the fair held in 1957 and that they could grant prizes nearly amounting to $10,000.[101]

Across the nation, hundreds of communities held science fairs, many of which drew widespread public interest. In 1947, 15,000 people attended Rhode Island's science fair in Providence, where 576 student exhibits were on display. The *Providence Journal* publicized and financed the event. In Pittsburgh, the Buhl Planetarium and Institute of Popular Science had begun hosting annual science fairs in 1940. By 1949, it secured college scholarships from local institutions of higher education and special awards from local industries for student winners. St. Louis inaugurated its annual science fair in 1948, and in the following year, the event featured over 1,000 student exhibits. Its chairman, high school science teacher Norman R. D. Jones, acknowledged the support of the *Saint Louis Star-Times*, and he cited fairs in Providence, Washington, DC, and Pittsburgh as valuable precedents. Teachers and students from St. Joseph Central High School in Missouri visited Kansas City's science fair in 1952.

In the following year, nearly half of the 304 students taking science classes at the school entered projects in their school fair.[102]

Impressive numbers of student exhibitors attracted overflow crowds at fairs in the 1950s. The Keene High School Science Fair in New Hampshire, which had begun in 1935, could no longer contain exhibits and visitors to the school auditorium. By 1956, the event spread to the school's classrooms, gymnasium, and science laboratories. In Westchester County, New York, 12,000 people braved a snowstorm in 1957 to view more than 1,000 exhibits at its science fair. The 1947 Lehigh Valley Science Fair in Allentown, Pennsylvania, had attracted 100 students to view 14 projects from seven schools. By 1958, 30,000 visitors viewed 822 exhibits selected from a pool of 8,087 student projects at 28 preliminary school fairs. In Pittsburgh, the Buhl Planetarium Science Fair had drawn 129 student exhibitors from 34 schools in 1944; by 1956, these numbers reached 847 and 189, respectively. Communities in 46 of the 48 states reported science fair activities in 1956. St. Louis, New York City, and Kansas City led the way with thousands of student exhibitors. Six state and regional fairs each drew participants from more than 100 schools. Although most fairs averaged roughly 2,000 visitors, some far exceeded that number. The San Francisco Bay Area Science Fair drew 58,000 people in 1956, and was followed by 40,000 attending the Topeka Regional Science Fair, and 30,000 visiting the Greater St. Louis Science Fair. Science Service estimated that 1,500,000 Americans viewed 187,000 science fair exhibits by elementary and secondary school students in the 1955–1956 school year.[103]

As a culminating event for this popular movement, Science Service inaugurated the National Science Fair in May 1950. Student finalists from 13 local and regional fairs presented their projects to the public at the Franklin Institute in Philadelphia and competed for $1,000 in prize money to fund their future research needs. By 1957, this annual event had grown to include 233 student finalists from 123 local, regional, and state fairs who showcased their projects in Los Angeles.[104] Watson Davis believed that the National Science Fair complemented Science Service's sponsorship of clubs and the talent search by promoting greater public awareness and appreciation of scientific research. It could also orient the hundreds of local and regional fairs to national issues and "help replenish the nation's inadequate supply of scientists."[105]

Science Service encouraged newspapers to fund local fairs and support the student winners' trips to the National Science Fair.[106] "If our hometown kids show us how exciting, interesting, and understandable science is," Margaret Patterson argued in 1950, "then it becomes a newsworthy and enjoyable way to keep up with fast-moving modern science."[107] Newspapers could also help to cultivate scientific talent. Science Service furnished them with an urgent political rationale: "There is no greater need today, in this time of the nation's peril, than for trained young men and women in scientific, engineering and technological fields." Fairs would educate those students who would "build the weapons for our defense" and serve as "the front line soldiers in the world-wide war on malnutrition and disease, which breeds military war."[108] Many newspapers accepted this invitation. The editors of the *Lafayette Journal and Courier* in Indiana explained that the expense was well justified: "If we may help to encourage more and better scientists, we will feel well repaid."[109] Editors of the *Midland Daily News* in Michigan similarly declared that "a constant crop of young scientists will be the key to the future strength of the United States." The managing editor of the *Providence Journal and Evening Bulletin* believed that sponsoring the fair constituted "one of the best promotion devices for getting and holding public approval."[110] By 1958, 99 newspapers helped to organize local science fairs and covered expenses for the winners' trips to the ninth annual National Science Fair in Flint, Michigan.[111] Other groups, including state academies of science and industrial organizations, furnished silver medals and "wish awards" for equipment to further students' research. Through these various sponsoring agencies, tens of thousands of high school students received support and incentives each year to participate in local, regional, and national science fair competitions.[112]

As with the talent search, National Science Fair organizers hoped that participants would gain further inspiration by touring research facilities and associating with some of the nation's most distinguished scientists. Students at the 1951 fair in St. Louis, for instance, received career advice from five Nobel Laureates, including Washington University's Chancellor Arthur Compton. When the finalists gathered in Washington, DC, in May 1952, they met at the White House with the First Lady, toured the National Bureau of Standards, and were shown artifacts at the Smithsonian Institution not open to the general public. Students attending the 1953 National Science Fair in Tennessee enjoyed special access to the Oak Ridge National Laboratory for atomic research. In 1957, chemist and Nobel Laureate Glenn Seaborg advised the 233 fair participants to pursue doctoral degrees and learn the Russian language.[113] Science Service also scheduled tours of the leading research universities in Los Angeles: "[To] glimpse such dreamed-of wonders as a synchrotron, a hypersonic

Figure 5.5 Science Clubs of America brochure invites newspapers to sponsor science fairs, 1950.

wind tunnel, a 'cobalt bomb' for cancer treatment, and full scale college laboratories."[114]

According to fair organizers Margaret Patterson and Joseph Kraus, "practically every science fair exhibitor has the opportunity of going to college on a scholarship when he graduates from high school."[115] Achieving distinction sometimes led to summer employment at various university, industrial, and government laboratories. Stephen Caine of Shreveport, Louisiana, who had won a fourth place prize at the 1957 fair for designing a corrosion control mechanism, was promptly hired by the Texas Eastern Transmission Corporation to fortify its 6,000 miles of buried steel gas transmission pipelines. Joel Frederic Lubar, who exhibited his homemade telescope at the 1955 fair in Cleveland, subsequently ground and calibrated a lens for the Cumberland Optical Company in Maryland. Suzan Lynn Hopkins, whose exhibit of the effects of an antibiotic on the digestive system of earthworms earned a prize, gained summer employment at the Infectious Disease Laboratories at the University of Iowa and then the Eli Lilly Company to develop a purification procedure for new antibiotics. With these additional incentives, fair organizers believed, participants could envision and pursue productive scientific careers.[116]

Some contended that these public displays of science strengthened democratic citizenship. According to Watson Davis, improved understanding of scientific methods of investigation would lead more Americans to seek and critically evaluate relevant information about political or social issues. "If we are confident that there can be a scientific democracy," he argued, "we must be confident that the people in general given the facts and the supposed conclusions will make the right decision."[117] The National Science Foundation's director, Alan T. Waterman, informed fair participants in 1952 that "the strength of science comes from its fundamentally democratic nature," because anyone could conduct a scientific investigation. Furthermore, the results transcended personal opinion or preference. "Each new finding must meet the challenge of the host of scientific workers everywhere," Waterman argued: "It is thus that the framework of science is built upon the solid foundation of confirmation in the face of healthy skepticism."[118]

More frequently, participants were urged to appreciate the nation's apparent technological deficiency in an era of tense international relations. As the Korean War raged overseas and Congress interrogated suspected communists at home, the chairman of the National Security Resources Board, Stuart Symington, praised science fairs for securing the United States. "Scientific and technical know-how have made this nation a leader among nations, and will keep it so," Symington eclared in 1951: "With this sort of watchful leadership America will never be caught technically unprepared."[119] In the nation's capital a year later, finalists attended a luncheon on "Science Manpower" with representatives from the US Department of Defense, the Office of Naval Research, and the National Science Foundation. Davis promoted the 1954 National Science Fair on his radio program by hosting Howard Meyerhoff, president of the Scientific Manpower Commission, who warned listeners of a pressing national shortage of scientists, engineers, and technicians.[120] Local fairs conveyed similar political messages as well. The American Institute, which had once sought to foster students' appreciation of nature, hosted fairs in the 1950s to meet "an urgent need for increasing our technical manpower to insure the safety and welfare of our nation."[121] The Cold War had oriented science fairs to a quest for future scientists who would help fortify the United States.

Shortly following the Soviet Union's launching of *Sputnik* in the fall of 1957 and the domestic panic that ensued, the US military began issuing special awards at the National Science Fair. Judges appointed by the Navy selected five finalists, along with more than 100 other students from regionally and locally affiliated fairs, for a five-day journey on its fleet at sea. Similarly, the Army awarded three students "whose exhibits are in the specialized areas of missiles, satellites, electronics, electronic calculators, mathematics, high and low temperatures, instrumentation, meteorology and medicine." Recipients won tours of the Jet Propulsion Laboratories in Pasadena, the Army Ballistic Missile Agency in Huntsville, or the Army Medical Center in the nation's capital. The Air Force selected two exhibits—one on air power and the other about space exploration—to include at its Airpower Panorama in Dallas in the fall of 1958.[122] As historian Andrew Hartman has shown, the Cold War increasingly oriented American education to military and industrial concerns in the late 1940s and 1950s. Students participating in the National Science Fair were similarly encouraged to appreciate these sorts of applications for their own scientific projects.[123]

In the postwar era, Watson Davis simultaneously proposed that science fairs strengthened democratic citizenship by cultivating powers of rational thinking, while rewarding the most promising young scientists who would help defend the nation. In this respect, Davis exemplified what Hartman has characterized as American educators' emerging belief that "meritocracy was at one with democracy."[124] Speaking before the American Society for Engineering Education in 1955, Davis praised the collaboration of educators, scientists, institutions of higher education, industries, and newspapers for supporting the National Science Fair in its first six years. Enlisting Cold War rhetoric, he declared that unlike their international rivals, Americans allowed students the freedom to select their academic and professional pursuits. Student choice through the science extracurriculum would satisfy the United States' technological demands "without resorting to the methods that we are confident will eventually ruin the fruitfulness of Soviet technology."[125] Davis erroneously assumed, however, that Science Service's competitions for American youth reflected or facilitated equal educational opportunities.

"A Highly Selected Strain of Guinea Pigs"

For the thousands of American high school seniors who competed in the Science Talent Search each year, the outcomes could determine whether they furthered their formal education and pursued scientific careers. The competition's organizers pursued meritocratic ideals in claiming that they searched indiscriminately for the nation's future scientists. As discussed in the previous chapter, educational psychologists Harold Edgerton and Steuart Henderson Britt devised the criteria for selecting 40 winners and 260 honorable mentions from the pool of competitors. Entrants' names and places of residence were withheld until after a panel of judges had determined the pool of winners and honorable mentions. Only a student's sex was made known throughout the process of evaluation. Far more students solicited application materials than those who submitted completed ones. In the inaugural competition, for instance, over 10,000 students requested application materials, but only 3,175 completed them. The task of taking the examination, submitting high school records, securing teacher recommendations, and composing the essay effectively comprised the first hurdle.[126]

Edgerton and Britt emphasized that the most valid criteria must be used in the earliest stages of evaluation. Judges therefore began by scoring the science aptitude examination, proceeding with the student's high school academic record, and concluding with the teacher recommendations to whittle the pool of candidates to 300. Only at this point did judges evaluate the 1,000-word "My Scientific Project" essay to choose 40 finalists.[127] If demonstrated proficiency in experimental or laboratory work constituted the most significant predictor of a student's scientific contributions, however, then this sequence may have led judges to overlook some of the most scientifically gifted youth. Along these lines, several educators questioned Edgerton and Britt's methods of selection. Some argued that essays and interviews were better indicators of an entrant's scientific promise than a science aptitude examination that did not gauge persistence or motivation to succeed. Others worried that those whose native language was not English encountered unnecessary disadvantages.[128] Meanwhile, a series of follow-up studies gauged the talent search's effectiveness in predicting successful scientific careers. Edgerton exhorted former winners to report on their subsequent activities and achievements as a service to science and society. "Each of you," he explained in 1948, "is a random sample of a highly selected strain of guinea pigs."[129]

Edgerton's initial findings were encouraging. Talent search finalists pursued science majors in college and entered scientific professions at an exceptionally high rate. For example, 251 of 258 former winners (96.9 percent) reported in 1949 that they had chosen a scientific field.[130] A grant from the National Science Foundation in 1957 allowed Edgerton to discover that all 80 of the 1942 and 1943 talent search winners had entered and remained in scientific careers. According to Watson Davis, who had helped secure the grant, such evaluative studies were critical. "Much more money must be devoted to such studies if this nation is to meet the challenge of the Communist powers," he declared: "Many of the most capable young people are still being wasted because of the lack of early identification and proper guidance of their development."[131]

The talent search nonetheless reflected and even exacerbated existing inequalities in American secondary education. Notably, racial minorities were severely underrepresented among the 40 winners and 260 honorable mentions each year from 1942 to 1958.[132] Edgerton and Britt found that only one African–American student was represented among the 220 male winners and honorable mentions (0.4 percent) in the inaugural competition; 21 of the 1,786 males (1.2 percent) who completed applications but did not win were African–American. Edgerton and Britt referenced educational psychologist Lewis Terman's claims that fewer gifted African–American youth existed proportionally to the larger population than for Whites, although they cautioned that intellectual differences along racial lines were inconclusive. Unlike Terman, however, Edgerton and Britt proposed that unequal educational opportunities played a role: "Since white pupils have greater odds of achieving senior class status in high school, they will be 'over-represented' in the whole mass of contestants in a search such as this."[133] Educational attainment, in other words, determined who could enter the talent search in the first place.[134] Despite Science Service's attempts to solicit entrants indiscriminately and to evaluate students' applications anonymously, only three African–American students appear to have been among the 680 talent search winners from 1942 to 1958.[135]

Unequal gender patterns also prevailed. Girls typically comprised from 25 to 30 percent of all applicants. Far fewer girls than boys were among the 40 talent search winners each year, as the proportion

Table 5.1 Science Talent Search winners by gender, 1942–1958

Year	Boys	Girls	Total	% Girls
1942	31	9	40	22.5
1943	29	11	40	27.5
1944	28	12	40	30.0
1945	30	10	40	25.0
1946	30	10	40	25.0
1947	30	10	40	25.0
1948	32	8	40	20.0
1949	31	9	40	22.5
1950	32	8	40	20.0
1951	29	11	40	27.5
1952	31	9	40	22.5
1953	31	9	40	22.5
1954	32	8	40	20.0
1955	32	8	40	20.0
1956	32	8	40	20.0
1957	31	9	40	22.5
1958	32	8	40	20.0
Total	523	157	680	23.1

Source: Permission granted by the Smithsonian Institution Archives.

of female entrants among all contestants determined the number of awards girls would receive. While Davis and Edgerton believed that this policy ensured that some girls would receive awards, it also limited how many girls could secure a place among the 40 winners and 260 honorable mentions. Citing the salience of "environmental and cultural factors," Edgerton and Britt called on educators to enact compensatory measures: "Greater attention in the primary and secondary schools to scientific training for American girls."[136] Table 5.1 reveals that a mere 157 of the 680 (23.1 percent) talent search winners from 1942 to 1958 were girls. Similarly, only 509 of the 2,080 (24.5 percent) of those receiving honorable mentions for select years between 1942 and 1958 were girls.[137]

Some female talent search winners subsequently reflected on incentives they encountered in pursuing scientific careers and how they attempted to balance their personal relationships with professional ambitions. Constance Sawyer Warwick reported in 1948 that she and her husband were pursuing doctoral degrees in astronomy at Harvard and Radcliffe and that they got "mixed up in housework and each other's homework."[138] Margaret Joan Hodgson announced her engagement to a physical chemist in 1948. Both hoped to work at the Oak Ridge Institute of Nuclear Research upon marriage.[139]

Others proudly conveyed their recent scientific achievements. Mary Ann Williams reported in 1949 that she was about to publish her first scientific paper as a coauthor: "It really gives me a sense of some sort of accomplishment and also a nice reward for the guinea pig tending which I have been doing for the past year and a half."[140] Meanwhile, Jean Towle conveyed optimism about balancing her domestic and professional duties: "This business of combining two careers—marriage and chemistry—is interesting, entertaining, and above all, time consuming." Towle reported working for Sinclair Research Laboratories and having "the honor of being the only female in the labs with my name on the lab door."[141] These sorts of testimonies suggest that some female talent search winners were fulfilling their scientific ambitions. Noting that a greater proportion of female talent search honorees went on to earn bachelor's and higher degrees than female non-honorees, Edgerton credited the talent search for bringing "confirmation of their own worth and ability."[142]

Cultural norms about female domesticity nonetheless placed severe obstacles to women's career advancement in the sciences. Popular notions that women should aid in their husbands' career advancement by fulfilling domestic roles discouraged some girls from pursuing scientific professions. And although the number of American women working rose in the 1950s, most of these jobs were part-time and career opportunities remained slim, which contributed to women's exclusion from the scientific professions.[143] Edgerton acknowledged in his 1961 follow-up study of the first two talent search cohorts that "there is a persistent expression that marriage is of first importance to fulfillment as a woman." He also found that many of those who attempted to balance scientific work with domestic responsibilities faced significant challenges: "The science role sometimes erected an 'egghead' barrier between them and their neighboring homemakers."[144] For example, Elisabeth J. Foster had reported in 1948 that she was engaged to one of her former students at the University of Chicago. Foster anticipated that marriage would alter her priorities: "If I find that sweeping floors...and working at least part time...does not completely occupy my time, I may take courses toward a Ph.D."[145] Nancy Slaven, in her final semester of college as a chemistry major, looked forward to marrying an industrial engineer in Birmingham, Alabama. "Needless to say," Slaven reported, "my mind is in Alabama most of the time, and my chemistry is focused on cooking edible biscuits."[146]

Instances of outright gender discrimination also abounded. As historian Margaret Rossiter has demonstrated, antinepotism rules at many institutions of higher education and industrial organizations in the postwar era derailed the potential careers of thousands of American female scientists.[147] In 1949, for example, former talent search winner Elizabeth Lyle encountered employment barriers because of her marriage and suburban residence: "Because of the difficulty in reaching Pittsburgh and because of the non-employment policy for married women adopted by the Aluminum Co., I have turned into a housewife."[148] Furthermore, the top salaries earned in 1951 by the first four cohorts of male talent search winners were at minimum 20 percent higher than those earned by female winners. These sorts of factors may have contributed to the relatively small number of girls who entered the talent search each year.[149] More generally, the ratio of girls taking science classes relative to boys declined across the nation. The percentage of girls among high school physics students fell from 29 percent to 24.4 percent from 1949 to 1958. Similar patterns emerged in high school chemistry and biology classes: from 44.3 percent to 40.9 percent and 53.2 percent to 49.4 percent, respectively.[150] According to historian Steven Mintz, postwar-era cultural values discouraged some female students from distinguishing themselves academically, as "many girls considered intellectualism and popularity mutually exclusive."[151] Cultural expectations about reified gender roles and policies of institutional discrimination thus limited the career prospects of some talent search alumnae. These were realities that the talent search's methods of selection did not address.

Pronounced geographical inequalities became evident as well. Of all 680 talent search winners from 1942 to 1958, 227 attended high school in New York State (33.38 percent), while none resided in Alaska, Arkansas, Delaware, Hawaii, Nevada, South Carolina, or Vermont. Population differences did not solely determine these disparities. For instance, New York State enrolled 9.06 percent of the nation's public high school seniors in the 1949–1950 school year. While acknowledging possible fluctuations in this ratio over time, it is notable that one-third of all talent search finalists from 1942 to 1958 came from high schools in New York State—well above the 9.06 percent one might expect. Illinois, which had the second highest number of talent search finalists with 53 of the 680 (7.79 percent), enrolled 5.11 percent of the nation's public high school seniors in the 1949–1950 school year. Pennsylvania ranked fifth among all states with 34 students among the 680 (5.00 percent) talent search winners from 1942 to 1958. As Pennsylvania enrolled 7.64 percent of the nation's public high school seniors in 1949–1950, however, it produced fewer winners than one might expect. Southeastern and primarily rural states furnished talent search winners especially infrequently as well. The highest-ranking state from the southeast was Virginia, which produced four of the 680 (0.59 percent) talent search winners while enrolling 0.97 percent of the nation's public high school seniors in 1949–1950.[152]

Economic inequalities partly explain these demographic patterns. Although the United States enjoyed unprecedented prosperity in the postwar era, millions of Americans remained mired in poverty, the highest rates of which occurred in the southern regions and often manifested in unequal schooling opportunities. Roughly one-third of all American children lived at or near poverty levels.[153] Highly urbanized states, meanwhile, tended to produce more talent search winners than those with higher proportions of rural residents. For example, only one of the ten least urbanized states in 1950 (West Virginia) produced talent search winners from 1942–1958 at a rate exceeding its proportion of public high school seniors. Four of these ten predominantly rural states (Alaska, Arkansas, South Carolina, and Vermont) yielded no winners during this span. Meanwhile, four of the nine most urbanized states in 1950 (Connecticut, Illinois, New Jersey, and New York) furnished talent search winners from 1942 to 1958 at rates highly exceeding their proportion of public high school seniors. The District of Columbia, with an exclusively urban population, produced talent search winners at a rate exceeding more than four times its national percentage of high school seniors.[154]

Curricular inequalities among high schools accounted for some of these discrepancies. In 1948–1949, for example, only 18 states (including the District of Columbia) had any students enrolled in advanced chemistry, and only 20 states had students enrolled in advanced biology. States without students in these two courses also tended to have fewer enrolled in high school physics. In the fall of 1958, moreover, 13 percent of public high school students had no physics courses available; 8.3 percent attended public high schools without any chemistry courses. As the Science Talent Search's judges weighed factors including the amount of science courses taken, as well as the sophistication of a student's essay on an original science project, those who attended smaller high schools with limited science offerings encountered fewer opportunities to submit competitive applications.[155] By contrast, teachers at Forest Hills High School in New York City selected high-achieving students to enroll in a second year of biology and to pursue original research projects. The BHSS gave students firsthand experiences in developing laboratory techniques through "extensive and intensive" course offerings and a well-developed extracurriculum. New York City's Stuyvesant High

School, with an enrollment of over 3,000 boys, offered a wide range of advanced electives, including experimental physics, laboratory techniques, and electronics. Factors such as these can explain why larger high schools located in or near metropolitan areas dominated in the talent search.[156] In 1953, the median senior class size of 39 of the 40 winners was 299. Only five winners came from senior classes of less than 100 students in that year.[157]

Shortly following the Soviet Union's launching of *Sputnik*, and the subsequent National Defense Education Act, James Bryant Conant published *The American High School Today*. Conant argued that because the fate of the United States depended on the training of its future leaders, large comprehensive high schools were essential for establishing differentiated curricula including rigorous science and mathematics courses. He simultaneously called for the consolidation of small high schools, because "wide academic programs are not likely to be offered when the academically talented in a school are so few in number." In Conant's estimation, more than 70 percent of American high schools (enrolling 30 percent of the nation's seniors) were too small to allow for sufficiently advanced courses. Figures from the US Office of Education supported Conant's claims. In the fall of 1958, 98.1 percent of public high schools enrolling 500 or more students offered at least one course in chemistry, while only 42.5 percent of public high schools enrolling 100 or fewer students did so. Whereas 96.9 percent of high schools enrolling 500 or more students offered at least one course in physics, only 33.4 percent of those with 100 or fewer students did so.[158]

Conant's assessment, coupled with the Science Talent Search's selection criteria, can explain why an overwhelming proportion of winners and honorable mentions from 1942 to 1958 attended large high schools and tended to reside in heavily urbanized states in the Northeast, Midwest, and far West. Rural youth and those residing in the Southeast, Southwest, and Mountain regions were especially underrepresented. Many talent search finalists benefited from the availability of a fully developed and specialized science curriculum and laboratory facilities, coupled with science teachers actively promoting a stimulating environment for scientific inquiry and specialization. Ambiguities about the precise meaning of "scientific ability" and the extent to which it could be measured may have contributed to these trends as well.

At the same time, Science Service's criteria and methods of selection discriminated against those students attending smaller high schools. By considering a student's class rank relative to senior class size, in addition to the number and types of science courses taken, talent search judges favored those attending larger high schools. Furthermore, quotas for female entrants, while ostensibly intended to guarantee their inclusion at some level, simultaneously dictated that fewer girls would be honored than boys overall. Institutional and cultural barriers rendered it difficult for most aspiring girls to become professional scientists in the postwar era. Although African–Americans' high school attendance rose significantly in the mid-twentieth century, moreover, many continued to lack access to specialized curricula and laboratory facilities. Despite their quest to summon the brightest minds for national defense in an atomic age, the Science Talent Search's architects did little to challenge sexist and racist ideologies that informed much of American educational history. As they did not account for pervasive inequalities in secondary schooling, these educators ensured that this prestigious annual competition would not be meritocratic.

Conclusion

American science educators created science clubs and fairs in the 1920s and 1930s to prompt children's investigation of their natural and social environments. According to the pedagogical innovator Morris Meister, firsthand experiences in scientific inquiry could teach people how "to think more clearly, more reasonably, and more honestly about the problems of life."[1] State Academies of Science similarly encouraged interested high school students to devise and display original science projects. These educators also believed that supervised extracurricular activities could help to mitigate the potentially pernicious effects of increased leisure time for youth. Hundreds of science clubs emerged across the United States, and dozens of communities inaugurated annual science fairs, the most elaborate of which took place at the American Museum of Natural History in New York City. These clubs and public exhibitions were not intended primarily to train future scientists. Instead, they sought to equip the next generation of citizens with scientific methods and knowledge to improve American society. Despite these democratic aims, however, only a select portion of American youth participated in these programs.

World War II transformed the civic characteristics of these science education activities. Disagreements at the New York World's Fair over the selection and presentation of students' research had exposed conflicting notions about the primary value of science to society. The subsequent mobilization of schools for national defense fueled a widespread network of science clubs and a new talent search. The scientific acumen of intellectually gifted students would fortify the United States militarily and materially in wartime. In the ensuing atomic age and Cold War, science fairs, clubs, and talent searches maintained their meritocratic aims in the quest for national defense and domestic prosperity. Five months before the Soviet Union's launching of *Sputnik*, for example, President Dwight D. Eisenhower sent his congratulations to the students at the 1957 National Science Fair for helping "to insure the future strength of our national economy and the freedom for all."[2]

Yet there was a prevalent fear that the United States faced a deficit of scientific expertise in the midst of a technological and ideological battle with the Soviet Union. These concerns prompted some national leaders to recommend systemic reforms in American science education. The National Science Foundation's president, Alan T. Waterman, sought federal legislation to that effect and cautioned "that the Soviet Union is graduating almost twice as many technical specialists in certain fields as in the United States."[3] Although he did not prescribe federal oversight, President Eisenhower exhorted Americans to remedy "the failure of us in this country to give high priority enough to scientific education and to the place of science in our national life."[4] Evidence of this inadequacy could be found in some studies of scientifically talented high school youth who did not proceed to college. "The United States is wasting its intellectual resources at the rate of approximately 200,000 18-year olds a year," proclaimed the College Entrance Examination Board in 1956. Increasing nationally allocated college scholarships to deserving students, as the Science Talent Search had done, could help mend the leaks in the educational pipeline.[5]

Watson Davis, Harlow Shapley, Meister, and others frequently argued in the postwar era that these extracurricular programs would develop rational thought for democratic citizenship. At the same time, the technological manpower messages had gained prominence—well in advance of the national panic stemming from the Soviet Union's launching of *Sputnik* in October 1957. As a result, Davis could attempt to reassure his radio listeners in February 1958 that science fairs, clubs, and talent searches had been grooming the nation's scientific experts for many years. "Long before the sputniks and our Explorer went into orbits the young scientists of America were preparing for our scientific future," he announced: "Here is evidence that Soviet Russia is not ahead of us in the skill, enthusiasm and knowledge of young scientists."[6] World War II and the ensuing Cold War thus aligned science clubs, fairs, and talent searches with the nation's military and economic imperatives. Its sponsors claimed that these activities propelled the most deserving students to advanced study in high school and college to join the ranks of the technological elite.

Yet rural youth, girls, and African–Americans were perennially underrepresented. The Science Talent Search's selection criteria did little to compensate for enduring inequalities in secondary education or patterns of discrimination that may have prevented potentially deserving students from gaining distinction. Despite the articulation of national goals for science education in the 1940s and 1950s, American schools remained largely decentralized and dependent primarily on local resources. As a result, students attending high schools without laboratory facilities or a developed curriculum in the sciences had few incentives or opportunities to compete successfully. The establishment of a military–industrial complex in the United States created a persistent demand for warships, airplanes, and atomic weapons. By contrast, the manifestations of science surrounding most rural youth—agriculture and the life sciences—appeared to hold less currency in that political context. Meanwhile, most Americans did not expect girls to pursue scientific careers, especially if those aspirations compromised the prospects of marriage and raising a family. And while the high school attainment of African–Americans increased in the postwar era,[7] unequal curricular resources abounded, which limited their participation and achievement in these science education programs.

As historian John Carson has shown, many American educators in the early twentieth century embraced narrow notions of intelligence in an attempt "to unify the democratic and meritocratic." As a result, all students could compete "for limited educational resources and occupational opportunities in ways that could appear objective and fair even to those least successful in garnering rewards from the system."[8] In the 1940s and 1950s, the architects of the Science Talent Search consistently declared that their selection methods were impartial and valid. Inviting all high school seniors in the United States to enter and compete appeared to satisfy one democratic criterion, but inequalities in American secondary science education unduly thwarted the participation of some students. Furthermore, proclamations about the pressing need to develop scientific talent through a national system of clubs, fairs, and competitions increasingly overshadowed the civic benefits of widespread scientific literacy. Rewarding some high-achieving students with scholarships to further their education and gain access to networks of governmental, scientific, and corporate leaders satisfied one meritocratic criterion. The persistent underrepresentation of rural youth, girls, and African–Americans nonetheless reflected inequities in American science education and discriminatory aspects of popular culture. Even during the economically prosperous postwar era, millions of Americans were mired in severe poverty.[9] Some American children thus encountered few opportunities to become interested in science—much less to aspire to scientific careers. It remained to be seen whether the wave of federal educational legislation beginning with the National Defense Education Act of 1958 that sought to address these differences would be effective in creating greater equality of educational opportunity.[10]

Science education has remained a high national priority in the United States. In an era of global economic competition, American political leaders continue to praise winners of science fairs and talent searches as vital resources for national security and material abundance.[11] New public–private partnerships for improving science education, meanwhile, evoke comparable efforts from generations ago. In the late 1930s, the Westinghouse Electric and Manufacturing Company supported science clubs and allowed students to showcase their research at the World's Fair. Westinghouse's subsequent funding of the talent search aimed to assist the nation militarily while demonstrating the virtues of industrial capitalism. In 1997, the CBS Corporation acquired Westinghouse, which ended its longstanding patronage of American science education. After 76 companies expressed interest, Science Service announced that one of the world's leading manufacturers of computers, the Intel Corporation, would become the new sponsor. Intel had also begun funding the International Science and Engineering Fair (formerly the National Science Fair).[12]

Recent prescriptions for reforming science education have similarly reflected older initiatives. The American Association for the Advancement of Science's (AAAS) report from 1990, *Science for All Americans*, articulated a global agenda for science education. Arguing that social and scientific issues transcended political borders, it promoted widespread critical thinking so that people would "use scientific knowledge and ways of thinking for personal and social purposes." Rather than grooming a class of technological elites, the AAAS report searched for ways to develop scientific literacy among all people: "To participate thoughtfully with fellow citizens in building and protecting a society that is open, decent, and vital." The quest to cultivate "habits of mind" and "compassionate" citizens who would "participate thoughtfully" resembled the civic rhetoric of science educators who created science clubs and fairs in the 1920s and 1930s. Both the AAAS report and these earlier science educators eschewed vocational justifications for science education and envisioned a kind of mass scientific literacy for societal progress. Both also encountered obstacles in pursuing that ideal. Like the corporate and military goals for science education that began to displace civic purposes in the 1940s and 1950s, more recent calls to enlist science education to bolster the nation's economy and security have dwarfed the AAAS's advocacy for global participatory democracy.[13]

As prime examples of this shift, the National Science Board's (NSB) 2006 report, *America's Pressing Challenge—Building a Stronger Foundation*, and the National Academies' (NAS) 2007 report, *Rising Above the Gathering Storm*, warned of a looming shortage of trained scientists. The NSB communicated a clear sense of urgency: "We cannot wait for a new *Sputnik* episode to energize our population to rise to this challenge—we must recognize the existing crisis and take the necessary actions." This meant improving K-12 science and math education for creating "the intellectual capital necessary to ensure this future workforce." More widespread scientific literacy, meanwhile, would elicit greater public appreciation and tax support for new curricular initiatives.[14] The NAS likewise lamented that "the scientific and technological building blocks critical to our economic leadership are eroding at a time when many other nations are gathering strength," which would compromise good employment and affordable sources of energy for Americans. A quotation by Nobel Laureate Julius Axelrod featured at the outset—"Ninety-nine percent of the discoveries are made by one percent of the scientists"— conveyed a pressing need to identify and groom future scientific elites. Casting their proposed reform measures in science education as a sound economic investment for the nation, the authors admonished that without active interventions "we can expect to lose our privileged position."[15] Ominously, the NAS reported in 2010 that "with regard to sustained competitiveness... our nation's outlook has worsened" because of deepening debt and "little sign of improvement, particularly in mathematics and science" in American public schools.[16]

Both the NSB and NAS reports reflect some of the dominant justifications for science fairs and talent searches, particularly from the 1940s and 1950s. Their overarching economic concerns are reminiscent of Westinghouse's emphasis on the consumer applications of scientific research and science education to revive Americans' confidence in industry. Furthermore, the quest to locate scientific elites to fuel the United States' economic and political power resembles the founding of the talent search competition to cultivate expert leadership in wartime. Indeed, both the NSB's and NAS's characterization of the looming retirement of a generation of scientists as a national crisis evokes World War II (and later Cold War) calls for science education to remedy a deficit of technological expertise.

While more optimistic in its outlook, the Domestic Policy Council (DPC) of the Office of Science and Technology Policy's 2006 report, *American Competitiveness Initiative: Leading the World in Innovation*, similarly argued that the nation's economic strength depended upon the quality of American science education. It also searched for ways to groom expert scientific leadership while training a scientifically literate workforce. In declaring "we will prepare our citizens to compete more effectively in the global marketplace," the DPC likened the roles of citizens to producers and consumers.[17] This predominant focus on science education's national economic benefits evokes Westinghouse's approaches to presenting science education at the New York World's Fair and beyond. In addition, the DPC's proposed tax incentives to entice roughly 30,000 professional scientists and engineers from private industry to join an "adjunct teaching corps" were part of an attempt to elicit greater corporate involvement in science education.[18] In light of the conflicts between science educators and Westinghouse officials at the World's Fair over half a century ago, such a development could spark new sorts of disagreements about the desired content, methods, and purposes of science education.

In tracing the contested civic dimensions of science in the United States, historian Andrew Jewett has pointed to World War II and the Cold War as a pivotal turning point. Shedding the "deliberative idealism" promoted by John Dewey in the early twentieth century in which communities actively utilized scientific knowledge to inform their values, the societal benefits of science transformed to material comforts in a consumer economy and a powerful military.[19] The changing justifications for science fairs, clubs, and talent searches are emblematic of this broader shift. As the nation mobilized for war in the early 1940s, organized science activities for American youth increasingly focused on grooming the most talented students for expert leadership. In the midst of the world war and a subsequent atomic age, the quest to develop habits of active inquiry and rational thought for democratic citizens appeared less urgent. Ironically, as the United States emerged as the world's leading economic and political power, it placed increasing emphasis on expert knowledge and leadership to secure its privileged position. Science educators and policymakers alike came to assume that the vast majority of Americans had neither the interest nor the capacity of developing informed opinions about the nation's technological needs.

It comes as little surprise, then, that the American polity appears to lack what Kenneth Prewitt has termed "scientifically savvy citizens" who can envision the societal implications of technological knowledge in nuanced ways.[20] Meanwhile, few contemporary educators or policymakers champion science education as a means of cultivating active, rational, and empathetic citizens in a participatory democracy.

Without this civic justification, many Americans seem to have little motivation to follow science news or to formulate and express their views about how public policy should direct scientific research and *vice-versa*. Perhaps by reintroducing democratic purposes in science education that had once inspired science clubs and fairs, a wider segment of youth can actively engage in science and steer its societal consequences: not merely for academic achievement, military strength, and material prosperity, but for the health of American public life.

Notes

Introduction

1. Hugo Newman, "The Children's Science Fair and the School," in Morris Meister, *Children's Science Fair of the American Institute: A Project in Science Education* (New York: The American Institute and the American Museum of Natural History, 1932), 26.
2. W. Stuart Symington, "Aids National Security," *Science News Letter* 59 (April 7, 1951): 221.
3. For accounts of the Westinghouse Science Talent Search, see Tom K. Phares, *Seeking—and Finding—Science Talent: A 50-year History of the Westinghouse Science Talent Search* (Pittsburgh: Westinghouse Electric Corporation, 1990); Joseph Berger, *The Young Scientists: America's Future and the Winning of the Westinghouse* (Reading, MA: Addison-Wesley, 1994); Ron Cowen, "'Go for It, Kid.' Looking Back on Five Decades of the Science Talent Search," *Science News* 139 (February 23, 1991): 120–123.
4. Carl Kaestle, *Pillars of the Republic: Common Schools and American Society, 1780–1860* (New York: Hill & Wang, 1983), 75–103; David B. Tyack, *The One Best System: A History of American Urban Education* (Cambridge, MA: Harvard University Press, 1974); Jurgen Herbst, *The Once and Future School: Three Hundred and Fifty Years of American Secondary Education* (New York: Routledge, 1996); Michael McGerr, *A Fierce Discontent: The Rise and Fall of the Progressive Movement in America, 1870–1920* (New York: Free Press, 2003).
5. Conceptions of democratic government and citizenship vary according to the desired breadth and depth of participation by its members. In a minimal view of democracy, all citizens enjoy the right to vote regularly to elect rulers and then trust those leaders to act without oversight. Prizing political stability above all, minimal democracy does not guarantee majority representation or rational rule. In a pluralist democracy, citizens express their interests as political equals by lobbying or pressuring government—although not all citizens are guaranteed access to interest groups as channels of communication. In a deliberative or participatory democracy,

citizens enjoy equal opportunities to debate differing views about public issues rationally in search of a mutually acceptable outcome. Carl Cohen, "Democracy [Addendum]," in *Encyclopedia of Philosophy, Volume 2,* ed. Donald Borchert, 703–706 (Detroit: Macmillan Reference USA, 2006); David Owen, "Democracy," in *Political Concepts,* ed. Richard Bellamy and Andrew Mason, 105–117 (Manchester & New York: Manchester University Press, 2003); Andrew L. Creighton, "Democracy," in *Encyclopedia of Sociology, Volume 1,* ed. Edgar F. Borgatta and Rhonda J. V. Montgomery, 601–609 (New York: Macmillan Reference USA, 2001).
6. John Dewey, *Democracy and Education* (New York: Free Press, 1966), 87.
7. John Dewey, "My Pedagogic Creed," *School Journal* 54 (January 1897): 77–80.
8. Dewey, *Democracy and Education,* 99.
9. John Dewey, *Experience and Education* (New York: MacMillan, 1963), 51–72; Robert Westbrook, "Public Schooling and American Democracy," in *Democracy, Education, and the Schools,* ed. Roger Soder, 125–150 (San Francisco: Jossey-Bass Publishers, 1996). See also Andrew Jewett, "Science & the Promise of Democracy in America," *Daedalus* 132 (Fall 2003): 64–70.
10. In a meritocracy, members earn educational and occupational rewards on the basis of achievement or by possessing valued attributes. The most prominent variant of meritocracy prizes productivity and efficiency in matching rewards for particular accomplishments. Thomas B. Hoffer, "Meritocracy," in *Education and Sociology Encyclopedia,* ed. David L. Levinson, Peter W. Cookson, and Alan R. Sadovnik, 435–442 (New York: RoutledgeFalmer, 2002); Matt Cavanagh, *Against Equality of Opportunity* (Oxford: Clarendon Press, 2002), 33; Norman Daniels, "Merit and Meritocracy," *Philosophy and Public Affairs* 7 (Spring 1978): 206–223; John Katsillis and J. Michael Armer, "Education and Mobility," in *Encyclopedia of Sociology Volume 2,* ed. Edgar F. Borgatta and Rhonda J. V. Montgomery, 755–760 (New York: Macmillan Reference USA, 2001); Patrick Akard, "Social and Political Elites," in *Encyclopedia of Sociology Volume 4,* ed. Edgar F. Borgatta and Rhonda J. V. Montgomery, 2622–2630 (New York: Macmillan Reference USA, 2001); Stephen J. McNamee and Robert K. Miller, Jr., *The Meritocracy Myth* (Lanham: Rowman & Littlefield Publishers, 2004), 101.
11. John Carson, *The Measure of Merit: Talents, Intelligence, and Inequality in the French and American Republics, 1750–1940* (Princeton: Princeton University Press, 2007); Paula S. Fass, *Outside In: Minorities and the Transformation of American Education* (New York: Oxford University Press, 1989), 44–72.
12. Amy E. Slaton, *Race, Rigor, and Selectivity in U.S. Engineering: The History of an Occupational Color Line* (Cambridge, MA: Harvard University Press, 2010), vii–xi.

13. John L. Rudolph, *Scientists in the Classroom: The Cold War Reconstruction of American Science Education* (New York: Palgrave, 2002), 31.
14. Rudolph, *Scientists in the Classroom*; Kim Tolley, *The Science Education of American Girls: A Historical Perspective* (New York: RoutledgeFalmer, 2003); Sally Gregory Kohlstedt, *Teaching Children Science: Hands-on Nature Study in North America, 1890–1930* (Chicago: University of Chicago Press, 2010). See also David Donahue, "Serving Students, Science, or Society? The Secondary School Physics Curriculum in the United States, 1930–65," *History of Education Quarterly* 33 (Fall 1993): 321–352; Philip J. Pauly, "The Development of High School Biology: New York City, 1900–1925," *Isis* 82 (December 1991): 662–688; Sally Gregory Kohlstedt, "'A Better Crop of Boys and Girls': The School Gardening Movement, 1890–1920," *History of Education Quarterly* 48 (February 2008): 58–93.
15. Important exceptions include Gerard Giordano, *Wartime Schools: How World War II Changed American Education* (New York: Peter Lang, 2004), and Charles Dorn, *American Education, Democracy, and the Second World War* (New York: Palgrave Macmillan, 2007). See also Andrew Hartman, *Education and the Cold War: The Battle for the American School* (New York: Palgrave Macmillan, 2008); John L. Rudolph, "From World War to Woods Hole: The Use of Wartime Research Models for Curriculum Reform," *Teachers College Record* 104 (March 2002): 212–241; Sevan G. Terzian, "'Adventures in Science': Casting Scientifically Talented Youth as National Resources on American Radio, 1942–1958," *Paedagogica Historica* 44 (June 2008): 309–325; David Kaiser, "Cold War Requisitions, Scientific Manpower, and the Production of American Physicists during World War II," *Historical Studies in the Physical and Biological Sciences* 33 (1) (2002): 131–159;
16. Rudolph, *Scientists in the Classroom*; Wayne J. Urban, *More Than Science and Sputnik: The National Defense Education Act of 1958* (Tuscaloosa: University of Alabama Press, 2010); Barbara Barksdale Clowse, *Brainpower for the Cold War: The Sputnik Crisis and National Defense Education Act of 1958* (Westport, CT: Greenwood Press, 1981); Hugh Davis Graham, *The Uncertain Triumph: Federal Education Policy in the Kennedy and Johnson Years* (Chapel Hill: University of North Carolina Press, 1984).
17. On patterns of gender discrimination in the history of American science education, see Tolley, *The Science Education of American Girls*; Margaret W. Rossiter, *Women Scientists in America: Before Affirmative Action, 1940–1972* (Baltimore: Johns Hopkins University Press, 1995); Sevan G. Terzian, "*Science World*, High School Girls, and the Prospect of Scientific Careers, 1957–1963," *History of Education Quarterly* 46 (Spring 2006): 73–99. On patterns of racial discrimination, see Slaton, *Race, Rigor, and Selectivity in U.S. Engineering*.

This investigation of science fairs, clubs, and talent searches also interprets these educational programs for youth as part of a broader effort to cultivate popular understanding and appreciation of science. See Marcel Chotkowski LaFollette, *Science on the Air: Popularizers and Personalities on Radio and Early Television* (Chicago: University of Chicago Press, 2008); Jessica Wang, *American Science in an Age of Anxiety: Scientists, Anticommunism, and the Cold War* (Chapel Hill: University of North Carolina Press, 1999); Nancy Smith Midgette, *To Foster the Spirit of Professionalism: Southern Scientists and State Academies of Science* (Tuscaloosa: University of Alabama Press, 1991); Marcel C. LaFollette, *Making Science Our Own: Public Images of Science, 1910–1955* (Chicago: University of Chicago Press, 1990); Peter J. Kuznick, *Beyond the Laboratory: Scientists as Political Activists in 1930s America* (Chicago: University of Chicago Press, 1987); John C. Burnham, *How Superstition Won and Science Lost: Popularizing Science and Health in the United States* (New Brunswick, NJ: Rutgers University Press, 1987); Bruce V. Lewenstein, "'Public Understanding of Science' in America, 1945–1965" (unpublished diss., University of Pennsylvania, 1987).
18. In 2008, Science Service, Inc. was renamed Society for Science and the Public.

1 Origins of Science Clubs and Fairs

1. Morris Meister, "Science Work in the Speyer School," *General Science Quarterly* II (May 1918): 429–445. The quotation appears on page 429.
2. Ibid., 445.
3. Born in Gonietz, Poland (then Russia), in 1895, Meister immigrated with his family to the United States in 1902. He grew up on the Lower East Side of Manhattan and was selected at age 12 to attend Public School 62, one of New York City's first accelerated or magnet schools. His interest in science emerged because of his teacher, Patrick Keenan, and Hugo Newman, who had designed an innovative curriculum. Meister later attended the prestigious Townsend Harris High School and earned a B.S. degree from City College of New York in 1916. To help fund his studies at City College, he began teaching English to other immigrants. This experience, Meister later reflected, inspired him to combine the study of science with teaching: "'Working with those adults who groped to master a new means of expression interested me tremendously, confirmed me in the desire to make teaching my lifework.'" It also prompted his involvement in educational programs for youth in settlement houses and summer camps. Meister quoted in Burton R. Pollin, *Toward Excellence in Education: Writings in Honor of Dr. Morris Meister* (Mission, KS: Inter-Collegiate Press, 1966), 7–8. On

Meister's early influence, see the account of a junior high school in Pittsburgh: Frank W. Murphy, "Science Clubs That Work," *General Science Quarterly* 4 (January 1920): 330–334.
4. Morris Meister, "The Educational Value of Certain After-School Materials and Activities in Science," (PhD diss., Columbia University, 1921); Emanuel Harris and Julius Lerner, "Dr. Meister's Early Years," in *Toward Excellence in Education*, ed. Pollin, 108.
5. Morris Meister, "Managing a Science Club," *School Science and Mathematics* 13 (March 1923): 205–217. The quotation appears on page 205.
6. Meister, "The Educational Value of Certain After-School Materials and Activities in Science," 62–64.
7. Steven Mintz, *Huck's Raft: A History of American Childhood* (Cambridge, MA: Harvard University Press, 2004), 214–229; Paula Fass, *The Damned and the Beautiful: American Youth in the 1920s* (New York: Oxford University Press, 1977), 16–21, 50–51; Howard Chudacoff, *The Evolution of American Urban Society*, 2nd ed. (Englewood Cliffs, NJ: Prentice-Hall, 1981), 211–212; Edward A. Krug, *The Shaping of the American High School* (New York: Harper & Row, 1964), 388.
8. Meister, "The Educational Value of Certain After-School Materials and Activities in Science," 73.
9. Meister, "Managing a Science Club," 208.
10. Meister, "The Educational Value of Certain After-School Materials and Activities in Science," 153–155; Meister, "Science Work in the Speyer School," 429–430.
11. Meister, "The Educational Value of Certain After-School Materials and Activities in Science," 102.
12. Ibid., 101–103.
13. George E. DeBoer, *A History of Ideas in Science Education: Implications for Practice* (New York: Teachers College Press, 1991), 65–89; Krug, *The Shaping of the American High School*, 373–375.
14. Arthur Zilversmit, *Changing Schools: Progressive Education Theory and Practice, 1930–1960* (Chicago: University of Chicago Press, 1993), 13–18; Lawrence Cremin, *The Transformation of the School: Progressivism in American Education, 1876–1957* (New York: Vintage Books, 1961), 218–220, 234–235; John Dewey, "My Pedagogic Creed," *School Journal* 54 (January 1897): 77–80; John Dewey, *Democracy and Education* (New York: Free Press, 1966), 86–88. On Dewey's conceptions of the compatibility of scientific inquiry and democratic citizenship, see Robert Westbrook, *John Dewey and American Democracy* (Ithaca: Cornell University Press, 1991), 141–142, 149.
15. Dewey did not serve on Meister's doctoral committee at Teachers College, but Kilpatrick did. The other members included the

esteemed science educator, Otis W. Caldwell, Thomas H. Briggs, and John F. Woodhull.
16. John L. Rudolph, "Epistemology for the Masses: The Origins of 'The Scientific Method' in American Schools," *History of Education Quarterly* 43 (Fall 2005): 341–376. The quotation appears on page 375. See also John L. Rudolph, "Portraying Epistemology: School Science in Historical Context," *Science Education* 87 (January 2003): 64–79.
17. Meister founded the Bronx High School of Science in 1938. Meister, "Science Work in the Speyer School," 445.
18. Meister quoted in Pollin, *Writings in Honor of Dr. Morris Meister*, 8–9.
19. Morris Meister, "Guiding and Aiding the Pupil in His Project," *General Science Quarterly* III (May 1919): 209–215. The quotation appears on page 209.
20. Ibid., 209; Meister, "Science Work in the Speyer School," 437.
21. Meister, "Science Work in the Speyer School," 435; Meister, "Guiding and Aiding the Pupil in His Project," 210–215.
22. Meister, "The Educational Value of Certain After-School Materials and Activities in Science," 144–145.
23. Meister, "Managing a Science Club," 217.
24. Meister, "Science Work in the Speyer School," 441.
25. Meister, "The Educational Value of Certain After-School Materials and Activities in Science," 153–155.
26. Meister, "Science Work in the Speyer School," 440; Meister, "The Educational Value of Certain After-School Materials and Activities in Science," 156–170; Meister, "Managing a Science Club," 212–217.
27. Meister, "The Educational Value of Certain After-School Materials and Activities in Science," 174–175.
28. Thomas Wessel and Marilyn Wessel, *4-H: An American Idea, 1900–1980* (Chevy Chase, MD: National 4-H Council, 1982), 1–49; George E. Farrell, *Miscellaneous Circular No. 85: Boys' and Girls' 4-H Club Work under the Smith-Lever Act, 1914–1924* (Washington, DC: United States Department of Agriculture, 1926); Fannie W. Dunn, "The Place of the 4-H Club Work in the American System of Public Education," *Proceedings of the Sixty-Sixth Annual Meeting of the National Education Association* 66 (1928): 508–517; Mary Eva Duthie, "4-H Club Work in the Life of Rural Youth," (PhD diss., University of Wisconsin, 1936), 1–4. The quotation is from Farrell, *Miscellaneous Circular No. 85*, II.
29. L. A. Robinson, "The Physics Club in a Normal School," *School Science and Mathematics* 7 (June 1907): 461–462; J. Arthur Lewis, "Experiences with Science Clubs," *School Science and Mathematics* 23 (October 1923): 624–629.
30. Willard N. Clute, "The High School Botanical Club," *School Science and Mathematics* 12 (February 1912): 147–149.

31. Ethel Bush, "Organizing the Biology Class Into a Nature Study Club," *Science Education* 15 (November 1930): 48–53.
32. Raymond Lussenhop, "The Organization of a Science Club," *School Science and Mathematics* 24 (October 1924): 727–730.
33. Karl F. Oerlein, "Science Clubs for Service," *School Science and Mathematics* 31 (March 1931): 314–320.
34. Mary Elizabeth Pape, "The Science Club," *School Science and Mathematics* 26 (May 1926): 552–554; Clarence M. Pruitt, "Activities of Chemistry Clubs," *Journal of Chemical Education* 4 (August 1927): 1037–1042.
35. Pruitt, "Activities of Chemistry Clubs," 1037–1042; Donald W. Miller, "Suggested Programs for a Science Club," *Science Education* 14 (November 1929): 331–334; Bush, "Organizing the Biology Class Into a Nature Study Club," 48–53; Emily Eveleth Snyder, "Report on the Biology Club of the Little Falls High School," *School Science and Mathematics* 31 (January 1931): 32–33; Francis P. Frazier, "General Science Club Notes," *School Science and Mathematics* 31 (March 1931): 341–344; Sarah W. Branch, "Science Club; Raleigh, N. Car.," *School Science and Mathematics* 31 (February 1931): 231–232; Murphy, "Science Clubs That Work," 330–334; Guy M. Smith, "Science Clubs in the High School," *School Science and Mathematics* 25 (October 1925): 720–725.
36. From 1922 to 1928, the proportion of American high school students taking physics dropped from 9 percent to 7 percent. The proportion taking chemistry remained steady at 7 percent, while the proportion taking biology rose from 9 percent to 14 percent. Those taking general science remained at 18 percent. Edward A. Krug, *The Shaping of the American High School Volume 2, 1920–1941* (Madison: University of Wisconsin Press, 1972), 56.
37. These developments caught the attention of Paul P. Boyd, president of the Kentucky Academy of Science, who devoted his 1920 address to the question of "what new forms of scientific service" state academies should assume. Two of Boyd's recommendations targeted secondary schools: To foster the growth of science clubs and to improve high school science teaching. Jane Ellen Gindelberger Kurzhals, "A History of the Illinois Junior Academy of Science," (MA thesis: MacMurry College, 1956), 4; J. L. Pricer, "Illinois State Academy of Science," *Science* 51 (March 26, 1920): 327; Louis A. Astell, "How State Academies of Science May Encourage Scientific Endeavor among High-School Students," *Science* 71 (May 2, 1930): 447; A. R. Crook, "Council Meeting of the Illinois State Academy of Science," *Science* 52 (December 3, 1920): 533.
38. Philip J. Pauly, *Biologists and the Promise of American Life: From Meriwether Lewis to Alfred Kinsey* (Princeton: Princeton University Press, 2000), 175–180.

39. Kurzhals, "A History of the Illinois Junior Academy of Science"; Astell, "How State Academies of Science May Encourage Scientific Endeavor among High-School Students"; Paul P. Boyd, "The Future of the State Academy of Science," *Science* 51 (June 11, 1920): 575–580; George W. Hunter, "What a Science Club Can Do for a School," *School Science and Mathematics* 23 (December 1923): 817–820.
40. L. J. Thomas and Arthur C. Walton, "A Brief History of the Junior Academy of the Illinois State Academy of Science," December 27, 1929, from Records of the American Association of the Advancement of Science, Washington, DC (hereafter AAAS Records), T-4-4, Box 10, Folder "Academy Conferences, 1928–1941," 4–5; Kurzhals, "A History of the Illinois Junior Academy of Science," 5–6; H. Carl Oesterling, "The Illinois Junior Academy of Science," *School Science and Mathematics* 31 (April 1931): 461–463; Astell, "How State Academies of Science May Encourage Scientific Endeavor among High-School Students," 447; Louis A. Astell and S. Aleta McEvoy, "Report of the Committee on High-School Science Clubs," *Transactions of the Illinois State Academy of Science* 23, no. 1 (1930): 25–30.
41. Astell, "How State Academies of Science May Encourage Scientific Endeavor," 445–449; Louis A. Astell and Charles W. Odell, *High School Science Clubs, Bureau of Educational Research Bulletin No. 60* (Urbana: University of Illinois, 1932), 28. McEvoy delivered an invited address about the JAS in Illinois at the AAAS meeting in Cleveland in December 1930; Astell addressed Indiana's fledgling Junior Academy in 1931. On how professional educators often responded to growing school enrollments in the early twentieth century, see David B. Tyack, *The One Best System: A History of American Urban Education* (Cambridge, MA: Harvard University Press, 1974), 182–255.
42. Hazel Elisabeth Branch, "The Aims and Opportunities of the Junior Academy in Kansas," *Transactions of the Kansas Academy of Science* 34 (April 1931): 27–32.
43. George E. Johnson, "Report of the Secretary," *Transactions of the Kansas Academy of Science* (April 1932): 23–24.
44. Krug, *The Shaping of the American High School*, 374–375; A. C. Walton, "The Academy Conference, Report of the Des Moines Session," December 27, 1929, AAAS Records, T-4-4, Box 10, Folder "Academy Conferences, 1928–1941," 1–2; A. C. Walton, "The Des Moines Session of the Academy Conference," *Science* 71 (February 7, 1930): 147–148; S. W. Bilsing, "The Academy Conference Minutes of the New Orleans Session, December 28, 1931," February 1932, AAAS Records, T-4-4, Box 10, Folder "Academy Conferences, 1928–1941," 1–4; Otis W. Caldwell, "Achievements and Obligations of Modern Science," in *Science*

Remaking the World, ed. Otis W. Caldwell and Edwin E. Slosson, 1–12 (Garden City: Doubleday, Page & Company, 1924); Morris Meister, "Obituary: Otis William Caldwell, 1869–1947," *Science* 106 (December 12, 1947): 576–578; Sally Gregory Kohlstedt, Michael M. Sokal, and Bruce V. Lewenstein, *The Establishment of Science in America: 150 Years of the American Association for the Advancement of Science* (New Brunswick: Rutgers University Press, 1999), 84–85, 97.

45. Louis A. Astell, "Fostering Science Clubs in the High School," *Journal of Chemical Education* 6 (March 1929): 496–501 (the quotation appears on page 497); Louis A. Astell, "The Inspiration Which the Junior Academy of Science Has Brought to the High School Science Clubs in the State of Illinois," *School Science and Mathematics* 32 (October 1932): 748–757. See also Clyde L. Exelby and Lida Belle Gambill, *Science Club Manual* (Lansing, MI: National Club Manual, 1931).

46. Ronald C. Tobey, *The Ideology of National Science, 1919–1930* (Pittsburgh: University of Pittsburgh Press, 1971); Kohlstedt, Sokal, and Lewenstein, *The Establishment of Science in America*, 83–87.

47. Astell and Odell, *High School Science Clubs*, 7–28.

48. Edwin Forrest Murdock, "The American Institute," in *A Century of Industrial Progress*, ed. Federic William Wile (New York: Doubleday, Doran & Company, 1928), v.

49. For example, the American Institute convinced the New York State Legislature to sponsor a bill to construct a waterway linking the Allegheny River and Erie Canal, which was passed in 1836. In that same year, the American Institute helped shape some of the provisions of a national law regarding patents. Horace Greeley, who assumed the American Institute's presidency in 1866, spearheaded a movement within the state legislature to establish national museums in New York City with exhibits depicting Americans' technological achievements, the nation's natural resources, American art and culture, and facilities including a lecture hall, library, chemical laboratory, and workshops to foster public knowledge of these matters. Despite these heady aspirations, the financial panic of 1873 undermined these efforts.

50. Murdock, "The American Institute," vi–xvi; "Old Science Body Begins A New Life," *New York Times*, June 24, 1928, X10.

51. Edwin F. Murdock, *Ninety-fifth Annual Report of the American Institute of the City of New York For the Year Ending January 16th, 1924* (New York, 1924), American Institute of the City of New York for the Encouragement of Science and Invention Records, 1808–1983, housed at the New-York Historical Society in New York City (hereafter NYHS AIR), Box 422, Folder 8; Edwin F. Murdock, *American Institute of the City of New York Ninety-sixth*

Annual Report For the Year Ending January 21st, 1925 (New York, 1925), NYHS AIR, Box 417, Folder 2, 3–13; Edwin F. Murdock, *97th Annual Report of the American Institute* (New York, 1926), NYHS AIR, Box 468, 3–8; Edwin F. Murdock, *Ninety-eighth Annual Report of the American Institute of the City of New York For the Year Ending December 31, 1926* (New York, 1927), NYHS AIR, Box 422, Folder 4.

52. L. W. Hutchins to W. D. Hendry, March 19, 1927, NYHS AIR, Box 418, Folder 2, 1; "Leroy W. Hutchins, Safety Counselor," *New York Times*, June 22, 1946, 14.
53. "General Plan for a Series of Popular Scientific Demonstrations to be Given in New York City During the Winter of 1928 under the Auspices of The American Institute of the City of New York" [c. 1928], NYHS AIR, Box 266, Folder 9, 3.
54. Speakers included Oscar Riddle, who discussed "Sex Research," Frank A. Arnold, who addressed "Broadcasting from the Layman's Point of View," and R. C. Beadle, who spoke about "The Reign of King Coal." E. F. Murdock, "To The Members of the American Institute of the City of New York" April 15, 1927, NYHS AIR, Box 266, Folder 9, 1–3; L. W. Hutchins, "Committee on Activities Report," May 15, 1927, NYHS AIR, Box 266, Folder 9, 1–2.
55. E. F. Murdock, *Ninety-ninth Annual Report of the American Institute of the City of New York For the Year Ending December 31, 1927* (New York, 1928), NYHS AIR, Box 468, 3–10; "General Plan for a Series of Popular Scientific Demonstrations," 4–5.
56. Hutchins anticipated an operating cost of $100,000 for the new programs. See "Prospectus of Operations for 1928," NYHS AIR, Box 140, Folder 3, 1–2.
57. E. F. Murdock to Marjorie Coit, July 3, 1928, NYHS AIR, Box 140, Folder 9.
58. L. W. Hutchins to Berne A. Pyrke, July 2, 1928, NYHS AIR, Box 140, Folder 4, 1.
59. Pauly, *Biologists and the Promise of American Life*, 180–185.
60. Sally Gregory Kohlstedt, *Teaching Children Science: Hands-On Nature Study in North America, 1890–1930* (Chicago: University of Chicago Press, 2010), 60.
61. "The American Institute Children's Fair Arranged by School Nature League Sponsored by and Held at the American Museum of Natural History," [1928], NYHS AIR, Box 140, Folder 3, 2. On the school gardening movement, see Sally Gregory Kohlstedt, "'A Better Crop of Boys and Girls': The School Gardening Movement, 1890–1920," *History of Education Quarterly* 48 (February 2008): 58–93.
62. Van Evrie Kilpatrick to Marjorie C. Coit, July 12, 1928, NYHS AIR, Box 140, Folder 9; L. W. Hutchins to Berne A. Pyrke, July 12, 1928, NYHS AIR, Box 140, Folder 4; "Meeting of the Fair Committee," September 7, 1928, NYHS AIR, Box 140, Folder

9, 1–2; Kohlstedt, "'A Better Crop of Boys and Girls,'" 58–93; Kohlstedt, *Teaching Children Science*, 166–167; Bessie Hersh, "The School Nature League: Its History, Organization and Educational Contributions" (MA thesis: Cornell University, 1940).
63. "The American Institute Children's Fair arranged by the School Nature League Sponsored and Held at the American Museum of Natural History Education Hall," October 18, 1928, NYHS AIR, Box 141, Folder 3; "The American Institute Children's Fair Arranged by the School Nature League Sponsored by The American Museum of Natural History," [c. 1928], NYHS AIR, Box 141, Folder 1, 6–9; Pauly, *Biologists and the Promise of American Life*, 171–193.
64. "Meeting of the Fair Committee," October 9, 1928, NYHS AIR, Box 140, Folder 9; Karen A. Rader and Victoria E. M. Cain, "From Natural History to Science: Display and the Transformation of American Museums of Science and Nature," *Museum and Society* 6 (July 2008): 152–171; Steven Conn, *Do Museums Still Need Objects?* (Philadelphia: University of Pennsylvania Press, 2010), 147; Sally Gregory Kohlstedt, "'Thoughts in Things': Modernity, History, and North American Museums," *Isis* 96 (December 2005): 586–601.
65. "The American Institute Children's Fair," [October 1928], NYHS AIR, Box 140, Folder 4, 1.
66. "Model Farm Shown at Children's Fair," *New York Times*, October 18, 1928, 21.
67. "Children's Fair Arranged by School Nature League"; "Institute's Children's Fair Is 99th Annual," *The Centurian* 1 (October 1928), NYHS AIR, Box 101, Folder 8, 9; A. R. Mann, "Pioneering with Nature," October 20, 1928, NYHS AIR, Box 141, Folder 3, 15.
68. E. F. Murdock, "Report of the Board of Trustees," October 23, 1928, NYHS AIR, Box 140, Folder 3, 1–2; Adeline C. Coorsen to E. F. Murdock, November 10, 1928, NYHS AIR, Box 140, Folder 3, 2; George Weinberger (to School Nature League), [c. 1928], NYHS AIR, Box 141, Folder 1; E. F. Murdock, *One Hundredth Annual Report of the Board of Trustees of the American Institute of the City of New York For the Year Ending December 31, 1928* (New York, 1929), NYHS AIR, Box 279, Folder 3, 3–5.
69. "Deficit in Science Group," *New York Times*, February 3, 1929, 20.
70. "American Institute Fails to End Row," *New York Times*, October 13, 1929, 14; "Scientists in Fight to Rule Institute," *New York Times*, November 2, 1929, 19; "American Institute Faces Fight Tonight," *New York Times*, November 7, 1929, 26; "Institute Groups in Lively Battle," *New York Times*, November 8, 1929, 27; "Seek Compromise in Institute Fight," *New York Times*, December 28, 1929, 7; E. F. Murdock, *One Hundred-First Annual Report of the American Institute of the City of New York For the Year Ending*

December 31, 1929 (New York, 1930), NYHS AIR, Box 283, Folder 9, 1–9; "Progressives Win Institute Contest," *New York Times*, February 14, 1930, 16; "Asks Ban on Merging American Institute," *New York Times*, July 22, 1930, 17; "American Institute Merger Ban Denied," *New York Times*, July 23, 1930, 26; "American Institute Votes for Merger," *New York Times*, July 24, 1930, 40.

71. "Thousands Attend the Children's Fair," *New York Times*, October 12, 1929, 26.
72. Mary Taylor, "Teaching Science Through the Children's Fair," *American Childhood* 15 (March 1930): 6–9; "Scout Statue Presented," *New York Times*, October 17, 1929, 26; "The American Institute Children's Fair Arranged by School Nature League Sponsored by and Held at the American Museum of Natural History Education Hall, October 11 to 17, 1929," Science Service Records, Smithsonian Institution Archives, Washington, DC, RU 7091 (hereafter SIA RU 7091), Box 101, Folder 8.
73. Morris Meister's notes to Marjorie Coit, [October 1928–January 1929], NYHS AIR, Box 142, Folder 1; Marjorie C. Coit to Morris Meister, January 29, 1929, NYHS AIR, Box 142, Folder 1; Meister, "The Educational Value of Certain After-School Materials and Activities in Science," 174–175.
74. L. W. Hutchins to Teachers of Science in High Schools of New York City, [c. 1929], NYHS AIR, Box 141, Folder 4; "The American Institute Children's Fair," April 25, 1929, SIA RU 7091, Box 101, Folder 8, 1–7; L. W. Hutchins to Gerald S. Craig, October 26, 1929, NYHS AIR, Box 142, Folder 5, 5.
75. Kohlstedt, *Teaching Children Science*, 221–225.
76. "1929 Children's Fair" meeting minutes, [c. 1929], NYHS AIR, Box 141, Folder 1, 2.
77. "School Nature League Children's Fair Committee meeting minutes," January 14, 1930, NYHS AIR, Box 141, Folder 1.
78. Carl A. Jessen and Lester B. Herlihy, *Offerings and Registrations in High-School Subjects, Bulletin 1938, No. 6* (Washington, DC: United States Department of the Interior, 1938).
79. "1929 Children's Fair" meeting minutes, 1–2; "Minutes of the Meeting of Judges and Committees of the 1930 Children's Fair," January 27, 1931, NYHS AIR, Box 141, Folder 1, 2; "The American Institute Children's Fair Arranged by School Nature League Sponsored by and Held at The American Museum of Natural History Education Hall Dec. 4 to 10, 1930," NYHS AIR, Box 142, Folder 2, 1–20; "The American Institute Children's Science Fair Arranged by School Nature League Sponsored by and Held at The American Museum of Natural History Education Hall Dec. 3 to 9, 1931," NYHS AIR, Box 142, Folder 2, 1–23.
80. Reuben Peterson to A. Cressy Morrison, November 28, 1930, NYHS AIR, Box 143, Folder 9; Reuben Peterson to Otis W.

Caldwell, [n.d.], NYHS AIR, Box 143, Folder 9; Marjorie Coit, "Radio," [c. 1930], NYHS AIR, Box 143, Folder 9, 1–4; A. Cressy Morrison, "Your Child Goes to School," December 2, 1930, NYHS AIR, Box 143, Folder 9, 1–4.
81. Otis W. Caldwell, "What The Children's Fair Means," [December 3, 1930], NYHS AIR, Box 143, Folder 9, 1–3.
82. Marjorie C. Coit, *Projects in Science and Nature Study* (New York: Department of Education of the Museum [of Natural History], 1931), 5.
83. Ibid., 11–12.
84. Ibid., 27.
85. Ibid., 33.
86. Morris Meister, "Report on the 1931 Children's Science Fair," [c. 1931], NYHS AIR, Box 143, Folder 5, 1–3; L. W. Hutchins, "The Fourth American Institute Children's Science Fair December 3–9, 1931," [December 1931], NYHS AIR, Box 141, Folder 1, 1–11; "Children's Own Fair of Science Is Opened," *New York Times*, December 4, 1931, 25; "A Children's Fair," *New York Times*, December 5, 1931, 16; "Children's Own Exhibits of Scientific Devices Outweigh Adults' in Their Appeal to Youth," *New York Times*, December 13, 1931, E7; Morris Meister, *Children's Science Fair of the American Institute: A Project in Science Education* (New York: American Institute and the American Museum of Natural History, 1932), 26.
87. A. Cressy Morrison, *104th Annual Report The American Institute of the City of New York for the Year Ending December 31, 1931* (New York, 1932), NYHS AIR, Box 468, 3–13; "Entries for Children's Science Fair" November 18, 1931, NYHS AIR, Box 144, Folder 7, 1; Meister, *Children's Science Fair of the American Institute*, 29–30; George N. Carothers, "A Science Fair," *Texas Outlook* 15 (April 1931): 16–20. Local science fairs also began in Pawtucket, Rhode Island, in 1928, and Roswell, New Mexico, in 1930.
88. Meister contributed to the National Society for the Study of Education's yearbook in 1932, a volume that attacked nature study as sentimental and obsolete. *The Thirty-First Yearbook of the National Society for the Study of Education: Part I, A Program for Teaching Science* (Bloomington, IL: Public School Publishing Company, 1932), 13–40; Kim Tolley, *The Science Education of American Girls: A Historical Perspective* (London: RoutledgeFalmer, 2003), 192; Kohlstedt, *Teaching Children Science*, 224.
89. Florence Weller et al., "A Survey of the Present Status of Elementary Science," *Science Education* 17 (October 1933): 193–198.
90. Meister, *Children's Fair of the American Institute*, 10.
91. Meister also prepared an hour-long "treasure hunt" for teams of junior high school students "to bring home to the students the idea that basic scientific principles affecting our daily life, the

meaning of the exhibits, rather than appearance, is the important thing." American Institute of the City of New York Press Release, "'Treasure Hunt' at Children's Science Fair," December 8, 1931, NYHS AIR, Box 144, Folder 7, 7–11, 20–22.

2 Building a Network

1. David M. Kennedy, *Freedom from Fear: The American People in Depression and War, 1929–1945* (New York: Oxford University Press, 2005), 162–163; David Tyack, Robert Lowe, and Elisabeth Hansot, *Public Schools in Hard Times: The Great Depression and Recent Years* (Cambridge, MA: Harvard University Press, 1984), 37; Steven Mintz, *Huck's Raft: A History of American Childhood* (Cambridge, MA: Harvard University Press, 2004), 234–238; Howard Chudacoff, *The Evolution of American Urban Society*, 2nd Ed. (Englewood Cliffs, NJ: Prentice-Hall, 1981), 251; Diane Ravitch, *The Great School Wars: New York City, 1805–1973* (New York: Basic Books, 1974), 236; Frederick M. Binder and David M. Reiners, *All The Nations Under Heaven: An Ethnic and Racial History of New York City* (New York: Columbia University Press, 1995), 176–182.
2. "Report by Dr. Meister to Board of Trustees on Junior Science Activities," June 15, 1932, NYHS AIR, Box 150, Folder 2, 1–2; Morris Meister, "Proposed Program for a Junior Institute of the American Institute," [April 1932], NYHS AIR, Box 151, Folder 9, 1.
3. "Minutes of a Meeting Held at the Hotel Astor, May 21, 1932," NYHS AIR, Box 223, Folder 9, 1–3.
4. Most leaders of the American Institute remained convinced that its science education programs constituted a sound investment, despite the immediate financial costs. Alfred Knight to H. T. Newcomb, June 16, 1932, NYHS AIR, Box 277, Folder 10, 2; Alfred Knight to H. T. Newcomb, July 5, 1932, NYHS AIR, Box 277, Folder 10, 2; Alfred Knight to DeLisle Stewart, June 22, 1932, NYHS AIR, Box 277, Folder 10; H. T. Newcomb to Alfred Knight, June 29, 1932, NYHS AIR, Box 277, Folder 10; Alfred Knight to Charles N. Frey, June 21, 1932, NYHS AIR, Box 277, Folder 10; Charles Frey to Alfred Knight, June 24, 1932, NYHS AIR, Box 277, Folder 10; Alfred Knight to H. T. Newcomb, June 28, 1932, NYHS AIR, Box 277, Folder 10; Alfred Knight to Charles N. Frey, June 29, 1932, NYHS AIR, Box 277, Folder 10.
5. "Minutes of a Meeting of the Plan Committee for Developing a Program of Activities of the Junior Science Clubs Organization," June 9, 1932, NYHS AIR, Box 223, Folder 9, 6; "New York's Young Scientists to Have City Wide Organization," September 23, 1932, NYHS AIR, Box 152, Folder 7, 1–2.

6. Morris Meister, "Radio Talk—WNYC the Children's Science Fair and the Junior Science Clubs Movement," January 18, 1933, NYHS AIR, Box 153, Folder 8, 1–7.
7. Morris Meister, *Living in a World of Science: Water and Air* (New York: Charles Scribner's Sons, 1930), xv.
8. "Statement Prepared by Dr. Morris Meister on the American Institute Junior Science Clubs," [1933], NYHS AIR, Box 150, Folder 2, 1–6.
9. "Minutes Junior Science Clubs Plan Committee Meeting," June 12, 1933, NYHS AIR, Box 221, Folder 1, 1–8; The American Institute, "Young Scientists of Bronx Conduct Demonstration Meeting," March 31, 1933, NYHS AIR, Box 268, Folder 7, 1–3; Morris Meister, "The Junior Science Clubs: An American Institute Project in Science Education," December 29, 1933, NYHS AIR, Box 155, Folder 5, 11–13.
10. The American Institute, "'Young Scientists' Hold First Meeting," October 27, 1932, NYHS AIR, Box 152, Folder 7, 1–2; "Minutes of Plan Committee Meeting Junior Science Clubs," September 22, 1932, NYHS AIR, Box 221, Folder 1, 1–6; "Minutes of Plan Committee Meeting Junior Science Clubs," October 6, 1932, NYHS AIR, Box 221, Folder 1, 1–11; "Summary of Report on 1st Central Meeting," October 29, 1932, NYHS AIR, Box 223, Folder 12, 1–6.
11. The American Institute, "Seven Hundred Young Scientists Are Guests of Museums," March 25, 1933, NYHS AIR, Box 152, Folder 7, 1–3; The American Institute, "American Institute Junior Scientists to Be Guests of Two Museums," March 23, 1933, NYHS AIR, Box 268, Folder 7, 1–2. On other demonstration lectures, see The American Institute, "Junior Science Clubs Hold Demonstration Meeting," January 13, 1933, NYHS AIR, Box 268, Folder 7, 1–2.
12. Steven Conn, *Do Museums Still Need Objects?* (Philadelphia: University of Pennsylvania Press, 2010), 13, 139.
13. The American Institute, "Three Meetings to Be Held for New York's 'Young Scientists,'" November 2, 1933, NYHS AIR, Box 223, Folder 16, 1–2. On March 20, 1934, students attended one of five programs: Earth science, "microphotography," biology for senior high school students, physical science for senior high school students, and a general session for junior high school students. Each program featured two speakers, a professional scientist or a science teacher. "Central Meeting The American Institute-Junior Science Clubs Saturday Morning March 10 at 10:30 A.M.," *The March of Science* II (February 28, 1934): 2–3. See also "Two Big Meetings," *The March of Science* III (October 19, 1934): 2–3; The American Institute, "Junior Scientists to Hear Latest Developments in Study of Structure of Plant Cells," March 20, 1935, NYHS AIR, Box 169, Folder 5.

14. The American Institute—Junior Science Clubs, "Outline of Museum Courses March to May 1934," [1934], NYHS AIR, Box 223, Folder 16, 1–11.
15. "The Life of the World," *The March of Science* II (February 8, 1934): 2–3; "The American Institute and the Junior Science Clubs," *School and Society* 41 (February 9, 1935): 185–186; "Prospectus of Workshop Courses November 1934–January 1935," [1934], NYHS AIR, Box 417, Folder 5, 1–12; "The American Institute Student Science Clubs Workshop Courses," *The March of Science* III (February 25, 1935): 2–3.
16. The American Institute, "American Institute Announces Junior Science Workshop Courses," [1934], NYHS AIR, Box 417, Folder 5, 1; "Speakers Bureau," *The March of Science* II (February 28, 1934): 4; "New Aims for Workshop Courses," *The March of Science* III (October 4, 1934): 1; "Broaden Your Outlook with a Chemistry Course," *The March of Science* III (November 15, 1934): 2–3.
17. "Workshop Courses," *The March of Science* 4 (October 21, 1935): 1.
18. "Workshop Courses," *The March of Science* 4 (September 23, 1935): 4; The American Institute, "Program, 1935–1936," [1935], NYHS AIR, Box 468, [no folders], 29–42; "Be an Early Bird at the Workshop Courses," *The March of Science* 5 (October 19, 1936): 2–3; "Science Club News," *Amateur Scientist* II (November 1937): 20; Gustave Wiedeman, "[Report-Sponsor Course]," [1937], NYHS AIR, Box 417, Folder 21, 1–4; "Laboratory Course in Fundamental Techniques and Skills for Science-Club Sponsors," [1937], Library and Archives Division, Sen. John Heinz History Center, Records from the George Westinghouse Museum, 2010.0153 (hereafter JHHC GWM), Box 69, Folder "Junior Science Exhibit"; "Prospectus of Workshop Courses March-May 1938," 1938, JHHC GWM, Box 69, Folder "Junior Science Exhibit," 1–6.
19. "Minutes Plan Committee Meeting," January 19, 1933, NYHS AIR, Box 151, Folder 1, 7–9; "Minutes of Plan Committee Junior Science Clubs," January 22, 1933, NYHS AIR, Box 221, Folder 1, 6–7; "Program Science Congress May 20, 1933," NYHS AIR, Box 268, Folder 6, 1–7.
20. "Science Congress the American Institute—Junior Science Clubs Program," June 2, 1934, NYHS AIR, Box 142, Folder 2, 1–8; "Were You at the Science Congress?" *The Amateur Scientist* II (March 1938): 20; "Program the American Institute Science Congress and Christmas Lectures for Student Science Clubs," December 28–29, 1936, NYHS AIR, Box 417, Folder 1, 1–8; "Program of the Science Congress—American Institute Science and Engineering Clubs," December 29, 1938, NYHS AIR, Box 217, Folder 7, 1.
21. "Program Science Congress May 20, 1933," NYHS AIR, Box 268, Folder 6, 1–7; "Science Congress The American Institute—Junior

Science Clubs Program, June 2, 1934," 1-8; "Program The American Institute Science Congress and Christmas Lectures for Student Science Clubs to Be Held at the American Museum of Natural History, December 26-27th, 1935," NYHS AIR, Box 142, Folder 2, 1-8; "Program The American Institute Science Congress and Christmas Lectures for Student Science Clubs," December 28-29, 1936, 1-8; "Program Science Congress American Institute School Science Clubs at the American Museum of Natural History December 18, 1937," JHHC GWM, Box 69, Folder "Junior Science Exhibit"; "Program of the Science Congress—American Institute Science and Engineering Clubs," 2-4; "Minutes Plan Committee Meeting Junior Science Clubs," May 9, 1933, NYHS AIR, Box 221, Folder 1, 1-6; "Children Hold Science Congress," [1933], NYHS AIR, Box 152, Folder 7; "Brooklyn Children Take Part in Science Congress," May 20, 1933, NYHS AIR, Box 152, Folder 7, 1-2; "Queens Children Take Part in Science Congress," May 20, 1933, NYHS AIR, Box 152, Folder 7, 1-2; "The Science Congress," *The March of Science* 1 (March 20, 1933): 3; "Program of the Science Congress—American Institute Science and Engineering Clubs," December 29, 1938, 1; Kim Tolley, *The Science Education of American Girls: A Historical Perspective* (New York: Routledge Falmer, 2003), 169-176.
22. "Junior Science Clubs of The American Institute Holds Science Congress," June 2, 1934, NYHS AIR, Box 223, Folder 16, 1-2; Morris Meister, "Why A Science Congress?" *The March of Science* II (May 8, 1934): 1, 4; Morris Meister, "Zone Meetings for Our Clubs," *The March of Science* II (February 28, 1934): 1; Mintz, *Huck's Raft*, 253.
23. Starting in 1935, the Science Congress was held in conjunction with the Christmas Lectures. "Listen in on the American Institute Christmas Lectures for Junior Science Clubs," [December 1934], NYHS AIR, Box 417, Folder 13; "The American Institute Christmas Lectures for Junior Science Clubs," *The March of Science* III (December 15, 1934): 1-3; "The American Institute Honors Science Students," December 27, 1934, NYHS AIR, Box 417, Folder 13, 1-2; "The American Institute Presents the Science Congress, the Christmas Lectures," *The March of Science* 4 (November 9, 1935): 1; "Don't Forget the Science Congress: Entries Due November 21st," *The March of Science* 5 (November 13, 1936): 1; "The Christmas Lectures on the Frontiers of Science," *The March of Science* 5 (December 12, 1936): 1-4.
24. L.W. Hutchins, "Talk before Park-West Neighborhood Association," November 7, 1934, NYHS AIR, Box 168, Folder 12, 1-2.
25. Meister, "The Junior Science Clubs," 14-17; Morris Meister, "The Junior Science Clubs: An American Institute Project in Science Education," *Science Education* 18 (April 1934): 68-74; Morris

Meister, "Pupil Adventure in Science," *High Points in the Work of the High Schools of New York City* 18 (September 1936): 5–12.
26. Wendt had been a professor of chemistry at the University of Chicago before assuming the American Institute's directorship. "Accent on Science Program #20," December 21, 1937, NYHS AIR, Box 201, Folder 4, 3–4. See also "Embryo Scientists to Demonstrate at Science Congress," [December 24, 1935], NYHS AIR, Box 414, Folder 5, 1; Robert Littell, "Staying After School for Fun," *The Reader's Digest* 31 (October 1937): 14–16; "Program Science Congress American Institute School Science Clubs," December 18, 1937, JHHC GWM, Box 69, Folder, "Junior Science Exhibit," 3.
27. *106th Annual Report the American Institute of the City of New York for the Year Ending December 31, 1933* (New York, 1934), NYHS AIR, Box 468, 5–6; *108th Annual Report the American Institute of the City of New York for the Year Ending December 31, 1935* (New York, 1936), NYHS AIR Box 468, 3; "Minutes Junior Activities Committee for 1935–36," June 26, 1935, NYHS AIR, Box 416, Folder 12, 1–2; "The Student Science Clubs," *The American Institute of the City of New York Monthly Bulletin* 1 (October 1936): 15, NYHS AIR, Box 170, Folder 6.
28. "Police May Teach Science to Young," *New York City Post*, December 15, 1932; "Science Clubs for Children," *Atlantic City N.J. Union*, December 13, 1932, newspaper clippings in NYHS AIR, Box 465; Alfred Knight to Christopher G. Rouse, June 6, 1933, NYHS AIR, Box 277, Folder 10, 1–2; "Minutes Plan Committee Meeting," January 19, 1933, NYHS AIR, Box 151, Folder 1, 4–6.
29. The American Institute-Junior Science Clubs, "Program: Science Club Course for the Crime Prevention Bureau, April 26–June 14," NYHS AIR, Box 165, Folder 2; "Minutes Committee on Cooperation with the Crime Prevention Bureau," April 25, 1933, NYHS AIR, Box 162, Folder 30, 1.
30. Catherine Emig to L. W. Hutchins, May 2, 1933, NYHS AIR, Box 263, Folder 24, 1–2; "Report Meeting to Discuss Cooperation with the Crime Prevention Bureau," June 8, 1933, NYHS AIR, Box 268, Folder 6, 2; "Minutes Committee on Cooperation with the Crime Prevention Bureau," 2–4; Frances Flanagan and Catherine Emig to L. W. Hutchins, May 24, 1933, NYHS AIR, Box 263, Folder 24, 1–2; "In Unison Lies Strength," *The March of Science* II (November 28, 1933): 3; "The Science Club Corner," *The Science Leaflet* 7 (January 4, 1934): 20.
31. The American Institute hosted five science fairs for New York City youth from 1932 to 1937; no fair was held in 1934. "The American Institute Children's Science Fair," [May 1936], NYHS AIR, Box 185, Folder 5, 1–2. See also Karen A. Rader and Victoria E. M. Cain, "From Natural History to Science: Display and the

Transformation of American Museums of Science and Nature," *Museum and Society* 6 (July 2008): 152–171; Steven Conn, *Museums and American Intellectual Life, 1876–1926* (Chicago: University of Chicago Press, 1998), 247; Sally Gregory Kohlstedt, "'Thoughts in Things': Modernity, History, and North American Museums," *Isis* 96 (December 2005): 586–601.

32. Catherine Emig Malcolm, "The Children's Science Fair a Project in Science Education," [1932], NYHS AIR, Box 145, Folder 3, 1–3; "The Children's Science Fair 1932," *The March of Science* I (September 17, 1932): 1; "Annual Children's Science Fair of the American Institute," *School and Society* 38 (October 14, 1933): 497; The American Institute, [Press Release], [May 8, 1937], NYHS AIR, Box 191, Folder 10, 4.

33. No fair was held in 1934. For the five years that the fair was held from 1932 to 1937, an average of 493 student exhibits were on display in the Education Hall at the American Museum of Natural History. "The Fifth American Institute Children's Science Fair," 1932, NYHS AIR, Box 145, Folder 3, 9; "The Sixth American Institute Children's Science Fair," 1933, NYHS AIR, Box 150, Folder 6, 5; "The Seventh American Institute Children's Science Fair," [1935], NYHS AIR, Box 150, Folder 6, 9; "The Eighth American Institute Children's Science Fair," 1936, NYHS AIR, Box 183, Folder 1, 12; [About the Ninth Science Fair], 1937, NYHS AIR, Box 193, Folder 12.

34. "Children's Science Fair, 1932–1937: Distribution of Exhibits, Grades 7-8-9," NYHS AIR, Box 209, Folder 14; "Children's Science Fair, 1930–1937: Distribution of Grouped Exhibits, Grades 7-8-9," NYHS AIR, Box 209, Folder 14.

35. "Children's Science Fair, 1932–1937: Distribution of Exhibits, Grades 10-11-12," NYHS AIR, Box 209, Folder 14; "Children's Science Fair: Distribution of Grouped Enteries [sic], Grades 10-11-12," NYHS AIR, Box 209, Folder 14.

36. Tolley, *The Science Education of American Girls*, 158–196.

37. Paula S. Fass, *Outside In: Minorities and the Transformation of American Education* (New York: Oxford University Press, 1989), 79–91, 240–253.

38. Catherine Emig, "The Children's Science Fair Offers Children and Young Women a Place in Science—Outside the Usual Interests of Boys," [1932], NYHS AIR, Box 145, Folder 3, 1–2; "Bronx Teacher Sponsors Prize-Winning Exhibits at the American Institute Children's Science Fair," May 22, 1936, NYHS AIR, Box 185, Folder 1.

39. Emig, "The Children's Science Fair," 2–3.

40. Tolley, *The Science Education of American Girls*, 179–196. The quotations are from pages 191 and 192.

41. Agnes G. Kelly, "Points of a Good Exhibit Children's Science Fair," *The March of Science* I (November 9, 1932): 1.

42. L. W. Hutchins, "The Children's Science Fair," April 11, 1935, NYHS AIR, Box 168, Folder 2, 3.
43. Catherine Emig, "Things on the Fair Which Need Much Thought," April 23, 1935, NYHS AIR, Box 174, Folder 19 (emphasis original).
44. "The Fifth American Institute Children's Science Fair," 5; "The American Institute Children's Science Fair Sponsored by and Held at the American Museum of Natural History Education Hall, April 7–14, 1935," [1935], NYHS AIR, Box 142, Folder 2, 4; "Children's Science Fair of the American Institute Sponsored and Held at the American Museum of Natural History Education Hall, May 9–16, 1937," [1937], NYHS AIR, 192, Folder 9, 2.
45. "The American Institute Children's Science Fair—May 9th–16th, 1937, Instruction to Judges," [1937], NYHS AIR, Box 191, Folder 10; "The American Institute Children's Science Fair—May 9th–16th, 1937, Definition of Criteria for Judging," [1937], NYHS AIR, Box 192, Folder 8.
46. "Digest of Criticisms of Judging," [1937], NYHS AIR, Box 191, Folder 7, 2.
47. "Junior Activities of the American Institute," [1935], NYHS AIR, Box 416, Folder 2, 7; *107th Annual Report the American Institute of the City of New York for the Year Ending December 31, 1934*, (New York, 1935), NYHS AIR, Box 468, 4–5.
48. "The Fifth American Institute Children's Science Fair," 10; "The Sixth American Institute Children's Science Fair," 8; "The Seventh American Institute Children's Science Fair," 10; "The Eighth American Institute Children's Science Fair," 13; "Attendance at the Fair," [1937], NYHS AIR, Box 193, Folder 12.
49. *107th Annual Report the American Institute of the City of New York*, 4–5; "Attendance at the Fair," [1937]; *109th Annual Report the American Institute of the City of New York for the Year Ending December 31, 1936* (New York, 1937), NYHS AIR, Box 468, 3; Ravitch, *The Great School Wars*, 237–239.
50. The American Institute of the City of New York, "[Press Release,]" May 10, 1937, NYHS AIR, Box 193, Folder 6, 1.
51. "Scientists of the Future," July 20, 1937, NYHS AIR, Box 199, Folder 21, 4.
52. "Editorial," *Amateur Scientist* 1 (May 1937): 18.
53. Peter J. Kuznick, *Beyond the Laboratory: Scientists as Political Activists in 1930s America* (Chicago: University of Chicago Press, 1987), 38–70; *110th Annual Report the American Institute of the City of New York for the Year Ending December 31, 1937* (New York, 1938), NYHS AIR, Box 468, 4.
54. Organizations expressing interest included the New York State Department of Agriculture, New York teachers' organizations, and the Marine Biology Station at Woods Hole in Massachusetts.

"The Fifth American Institute Children's Fair," 10; "The American Institute of the City of New York," [1932], NYHS AIR, Box 150, Folder 6, 12; "The Sixth American Institute Children's Science Fair," [1933]; "The Seventh American Institute Children's Science Fair," [1935], NYHS AIR, Box 172, Folder 12, 9; Herbert Zim to E. J. Kelly, September 26, 1938, NYHS AIR, Box 209, Folder 14.
55. "The Seventh American Institute Children's Science Fair," 9; L. W. Hutchins to Michael H. Lucey, June 20, 1935, NYHS AIR, Box 173, Folder 10; Gerald Wendt form letter to students, May 15, 1937, NYHS AIR, Box 191, Folder 10; "Junior Science Fair—May 8–12," *The American Institute Monthly* 2 (May 1938): 7.
56. Otis W. Caldwell, "Science Essays by High School Pupils," *Science* 75 (April 8, 1932): 385–388 (the quotation appears on page 386).
57. Burton E. Livingston, "The Atlantic City Meeting of the American Association for the Advancement of Science and Associated Studies," *Science* 77 (February 3, 1933): 129; Louis A. Astell, "The Inspiration Which the Junior Academy of Science Has Brought to the High School Science Clubs in the State of Illinois," *School Science and Mathematics* 32 (October 1932): 748–757; Jane Ellen Gindelberger Kurzhals, "A History of the Illinois Junior Academy of Science" (MA thesis, MacMurry College, 1956), 8; Mintz, *Huck's Raft*, 241.
58. States with Junior Academies of Science in 1938 (with number of active clubs and estimated membership, respectively) included Alabama (27 and 700); Illinois (80 and 2,000); Indiana (41 and 1,140); Iowa (30 and 500); Kansas (21 and 500); Kentucky (31 and 800); Minnesota (7 and 100); Missouri (no information available); Nebraska (375 estimated membership); Oklahoma (12 and 360); Pennsylvania (34 and 2,000); St. Louis (separate from Missouri with 18 and 450); Texas (15 clubs); and West Virginia (38 and 800). [Junior Academies of Science], AAAS Archives, T-4-4, Box 10, Folder, "Academy Conference, 1928–1941," 1–24; "Out of the State," *News Notes* II (September 1937): 7–8.
59. S. W. Bilsing, "Science Clubs in Relation to State Academies of Science," *Science Education* 18 (October 1934): 162–167. See also M. M. Williams, "The Junior Academy of Science, Its Present Organization and Future Possibilities of Our State and Nation," *Teachers College Journal* 5 (September 1933): 148–152.
60. Otis W. Caldwell, "From the Viewpoint of the Interrelationship of National, State and Local Science Organizations and Clubs," *Science Education* 22 (February 1938): 70–71.
61. John C. Burnham, *How Superstition Won and Science Lost: Popularizing Science and Health in the United States* (New Brunswick: Rutgers University Press, 1987).
62. Karl F. Oerlein, "Junior Academies of Science," *Educational Outlook* 14 (November 1939): 9–20.

63. Howard E. Enders, "Introductory Remarks Concerning the Origin of the Junior Academy of Science and Its Relation to the Indiana Academy of Science," *Teachers College Journal* 5 (September 1933): 146–148.

64. Science fair organizers in Boston and Duluth, Minnesota, also asked the American Institute for guidance. "Children's Science Fair 1936 Sponsored by the Wollaston Mothers' Club and Held at the Unitarian Church April 24 & 25, 1936," NYHS AIR, Box 175, Folder 5; Alfred F. Nixon to Morris Meister, January 5, 1938, NYHS AIR, Box 175, Folder 5; Gerald Wendt to Alfred F. Nixon, January 14, 1938, NYHS AIR, Box 175, Folder 5; E. J. Kelly to Walter C. Alvarez, January 14, 1938, NYHS AIR, Box 204, Folder 13; I. B. Ham to the American Institute, May 13, 1938, NYHS AIR, Box 204, Folder 13; H. H. Sheldon to I. B. Ham, June 13, 1938, NYHS AIR, Box 204, Folder 13; Mrs. Zebulon Vance to the American Institute, October 18, 1937, NYHS AIR, Box 175, Folder 5.

65. Bertha E. Slye to Gerald Wendt, March 17, 1938, NYHS AIR, Box 175, Folder 5; Edwin W. Nelson to American Museum of Natural History, February 15, 1937, NYHS AIR, Box 175, Folder 5; Mabel Carson to the American Institute, March 20, 1937, NYHS AIR, Box 175, Folder 5; Bertha E. Slye to the American Institute, February 4, 1937, NYHS AIR, Box 175, Folder 5; E. J. Kelly to Bertha E. Slye, March 28, 1938, NYHS AIR, Box 175, Folder 5; "Hundreds See Science Fair," *Detroit Times* [1938], newspaper clipping in NYHS AIR, Box 175, Folder 5; Hyman H. Platt to [Catherine] Emig, June 27, 1934, NYHS AIR, Box 175, Folder 5; "Science Fair Elizabeth Peabody House, 357 Charles St. Boston, Mass. Feb. 26–27, 1938," NYHS AIR, Box 175, Folder 5; Earl M. Hauer to Catherine Emig, August 16, 1935, NYHS AIR, Box 175, Folder 5; Catherine Emig to Earl M. Hauer, September 3, 1935, NYHS AIR, Box 175, Folder 5.

66. Walter Kozovick to the American Institute, August 2, 1938, NYHS AIR, Box 204, Folder 13; Andrew C. Whyte to the American Institute, October 25, 1935, NYHS AIR, Box 186, Folder 9; Alexander Ganz to the American Institute of Student's Science Clubs, October 29, 1935, NYHS AIR, Box 186, Folder 9; Catherine Emig to Joseph A. Kuntz, December 1, 1936, NYHS AIR, Box 186, Folder 7; Bessie Schwartz to Merle M. Boyer, June 7, 1937, NYHS AIR, Box 175, Folder 5; Mrs. Hillyer Emily Brown to Catherine Emig, [1937], NYHS AIR, Box 175, Folder 5; "The American Institute Junior Science Clubs," March 16, 1938, NYHS AIR, Box 208, Folder 19, 2.

67. William B. Duryee, "The 1935 New Jersey Children's Science Fair," May 1935, NYHS AIR, Box 175, Folder 5, 5.

68. "The Children's Science Fair of the American Institute, New York," *School and Society* 41 (April 13, 1935): 502; Duryee, "The 1935 New Jersey Children's Science Fair," 5–21.
69. Duryee, "The 1935 New Jersey Children's Science Fair," 5–21.
70. Sarah Bent Ransom, "The Science Fair as an Aid to Project Teaching," *Science Education* 22 (March 1938): 138; Harry B. Weiss, "The New Jersey Science Fair, 1937," May 1937, NYHS AIR, Box 417, Folder 15, 1–23; H. B. Weiss, "Notice," November 29, 1937, NYHS AIR, Box 417, Folder 15; Maitland P. Simmons to the American Institute, December 14, 1937, NYHS AIR, Box 204, Folder 13; Gerald Wendt to Maitland P. Simmons, January 4, 1938, NYHS AIR, Box 204, Folder 13.
71. Edith R. Force, "Abstract of Report of the Director of the High School Relations Committee of the Oklahoma Academy of Science. December, 1935–December, 1936," *News Notes* II (January 1937), NYHS AIR, Box 328, Folder 7, 12; Edith R. Force, "Special Activities of Science Students," *Education* 56 (March 1936): 438–440; Otto M. Smith, "Greetings," *News Notes* I (November 1936), NYHS AIR, Box 328, Folder 7, 3–4; "Stillwater Meeting A Success," [c. 1937], NYHS AIR, Box 328, Folder 7.
72. Note attached to letter by Edith R. Force to Catherine Emig, November 23, 1936, NYHS AIR Box 328, Folder 7. In 1937, however, the Oklahoma Academy of Science dissolved its Committee on High School Relations, which prompted Force to confess to Catherine Emig that she was "ashamed for Oklahoma." Force concluded that national sponsorship of science education was necessary for yielding "science service in a democracy." Edith R. Force to the American Institute, August 10, 1936, NYHS AIR, Box 328, Folder 7; Catherine Emig to Edith R. Force, September 9, 1936, NYHS AIR, Box 328, Folder 7, 1–2; Edith R. Force to Catherine Emig, January 24, 1937, NYHS AIR, Box 328, Folder 7; "From Here and Yon," *News Notes* II (January 1937), NYHS AIR, Box 328, Folder 7, 10; E. R. F., [no title], *News Notes* II (March 1937), NYHS AIR, Box 328, Folder 7, 2; Edith R. Force to Catherine Emig, December 12, 1937, NYHS AIR, Box 328, Folder 7, 1–2; Catherine Emig to Edith R. Force, December 16, 1937, NYHS AIR, Box 328, Folder 7; Edith R. Force to Otis W. Caldwell, December 24, 1937, NYHS AIR, Box 417, Folder 17, 1–2; Edith R. Force, "The Need for a Twelve Year Science Program for American Public Schools," December 30, 1937, NYHS AIR, Box 417, Folder 17, 6.
73. "The Fifth American Institute Children's Science Fair," 10; "The Sixth American Institute Children's Science Fair," 9; "The Seventh American Institute Children's Science Fair," 8–9; "Junior Activities of the American Institute," [1935], NYHS AIR, Box 416, Folder

2, 2; "Suburban Club Members Allowed to Exhibit for First Time," *The March of Science* 4 (February 24, 1936): 1.

3 Showcasing Young Scientists at the New York World's Fair

1. "Youthful Scientists Open 'Laboratory' at N.Y. World's Fair," *Science Observer* 1 (May 1939): 1, 8.
2. Overall, roughly 45 million visitors attended the World's Fair from 1939 to 1940, and an estimated 10 million visited the building sponsored by Westinghouse. *The American Institute Science and Engineering Clubs' Exhibit at the World's Fair in Cooperation with the Board of Education of the City of New York, Westinghouse Building* (New York, 1939), NYHS AIR, Box 209, Folder 9.
3. Alfred Knight to H. H. Sheldon, November 3, 1935, NYHS AIR, Box 170, Folder 1; H. D. Lufkin to Alfred Knight, October 11, 1935, NYHS AIR, Box 170, Folder 1; H. D. Lufkin to Alfred Knight, December 10, 1935, NYHS AIR, Box 170, Folder 1; L. W. Hutchins to H. T. Newcomb, August 13, 1936, NYHS AIR, Box 170, Folder 2; Gerald Wendt to H. T. Newcomb, July 1, 1937, NYHS AIR, Box 178, Folder 4; Peter J. Kuznick, "Losing the World of Tomorrow: The Battle over the Presentation of Science at the 1939 New York World's Fair," *American Quarterly* 46 (September 1994): 341–373; Robert W. Rydell, "The Fan Dance of Science: American World's Fairs in the Great Depression," *Isis* 76 (December 1985): 525–542.
4. Gerald Wendt, "Science at the New York World's Fair, 1939," [1937], NYHS AIR, Box 176, Folder 3.
5. "Minutes of a meeting of the World's Fair Committee of the American Institute," July 13, 1937, NYHS AIR, Box 178, Folder 4.
6. "The American Institute of the City of New York World's Fair Committee," July 13, 1937, SIA RU 7091, Box 200, Folder 3; Watson Davis to Frederick A. Gutheim, July 14, 1937, SIA RU 7091, Box 200, Folder 3; Watson Davis, "Science in the New York World's Fair 1939," July 17, 1937, SIA RU 7091, Box 200, Folder 3.
7. Alan R. Ferguson to Gerald Wendt, August 10, 1937, NYHS AIR, Box 176, Folder 3.
8. The American Institute, "Plan for Junior Science Center at the New York World's Fair 1939," February 1938, NYHS AIR, Box 177, Folder 5, 1–26. All quotations are from pages 1–2.
9. The American Institute, "Minutes Student Science Clubs Executive Committee," November 14, 1934, NYHS AIR, Box 289, Folder 2, 1–2.
10. "The American Institute Student Science Clubs," [1937], NYHS AIR, Box 165, Folder 9, 1–12; Edith Force and Otto Smith, "News Notes," March 1937, NYHS AIR, Box 328, Folder 7, 1–3.

11. "School Children's Science Activities at the World's Fair," [March or April 1938], NYHS AIR, Box 180, Folder 6, 1–3.
12. "American Institute's World's Fair Committee meeting minutes," April 25, 1938, NYHS AIR, Box 179, Folder 7.
13. "Notes on a Meeting in Connection with the Expansion of the JSC," January 5, 1938, NYHS AIR, Box 408, Folder 7, 7.
14. "Notes on a Meeting for the Expansion of the JSC," December 28, 1937, NYHS AIR, Box 408, Folder 7, 5.
15. Ibid., 1–8.
16. "Preliminary Proposal for Extending Nationally the Junior Science Club Movement," [January 1938], NYHS AIR, Box 205, Folder 7, 1.
17. Gerald Wendt to Robert T. Pollock, February 24, 1938, NYHS AIR, Box 205, Folder 7, 1–2.
18. The American Institute, "A National Organization of Junior Science Clubs," February 1938, NYHS AIR, Box 408, Folder 7, 1–41; The American Institute, "Plan of Expansion for the American Institute Science and Engineering Clubs," [August 1938], NYHS AIR, Box 409, Folder 7, 1–7; The American Institute, "Plan for Promotion of the American Institute Science and Engineering Clubs," [August 1938], NYHS AIR, Box 409, Folder 7, 1–11; The American Institute, "Personnel Required for Direction of the American Institute Science and Engineering Clubs," [August 1938], NYHS AIR, Box 409, Folder 7, 1–13.
19. No specific evidence reveals exactly why Westinghouse decided to fund these science education programs. However, the company's general anxiety about its public image (especially relative to its main rival, General Electric) informed its plans for the World's Fair as a commercial endeavor above all. The quest to curry greater public favor, therefore, was likely to have motivated its sponsorship of the American Institute. *111th Annual Report the American Institute of the City of New York for the Year Ending December 31, 1938* (New York, 1939), SIA RU 7091, Box 205, Folder 6, 5–8; Stanley Edgar Hyman and St. Clair McKelway, "Onward & Upward with Business & Science: The Time Capsule," *The New Yorker* 29 (December 5, 1953): 196–203.
20. Charles A. Federer, Jr., "Nation-wide Junior Science Clubs," *Science* 88 (December 2, 1938): 526. See also The American Institute Press Release from October 21, 1938, NYHS AIR, Box 408, Folder 3; [Ruth Bayard], "Questions and Answers," October 24, 1938, NYHS AIR, Box 408, Folder 3.
21. The American Institute of the City of New York, *How to Organize a Science Club* (New York: The American Institute, 1938), 1–35. The quotation from Meister appears on page 3.
22. "A Word to Youth," *Science Observer* 1 (January 1939): 2.
23. "The American Institute Science and Engineering Clubs: Report on Nation-wide and Local Progress," November 21, 1938, NYHS

AIR, Box 409, Folder 2, I–VI; Robert T. Pollock, "Science Clubs for Nation's Youth," *Science Observer* 1 (December 1938): 1.

24. "Clubs Will Prepare World's Fair Exhibit," *Science Observer* 1 (January 1939): 1; "Replies to Flyer by State as of November 10, 1938," NYHS AIR, Box 409, Folder 2; "Addendum as of November 29, 1938," NYHS AIR, Box 409, Folder 2; "Recapitulation of Statistics as of Nov. 29, 1938," NYHS AIR, Box 409, Folder 2; "Junior Activities—General Report March 29, 1939," NYHS AIR, Box 409, Folder 2; "Proposals for Science Club Promotion at World's Fair," April 1, 1939, NYHS AIR, Box 409, Folder 2, 1–3; Charles A. Federer, Jr. to H. H. Sheldon, April 19, 1939, NYHS AIR, Box 409, Folder 2, 1–2; "National Clubs as of April 19th by States," [1939], NYHS AIR, Box 409, Folder 2, 1–2; "General Memorandum," May 31, 1939, NYHS AIR, Box 409, Folder 2; "First National Club Gets Certificate at a Formal Ceremony," *Science Observer* 1 (April 1939): 1.

25. "Science Clubs Supported by Westinghouse Company," *Science Observer* 1 (February 1939): 1.

26. Robert T. Pollock, "Science Clubs for Nation's Youth," *Science Observer* 1 (December 1938): 1. Upon acknowledging the receipt of the first $5,000 check from Westinghouse, Pollock even proposed that the American Institute's junior activities be renamed the "George Westinghouse Junior Science and Engineering Clubs of the American Institute" to elicit the "deep regard and high affection by all people in this nation for George Westinghouse." Robert T. Pollock to J. F. O'Brien, Esq., April 22, 1938, JHHC GWM, Box 69, Folder "Junior Science Exhibit."

27. Alexander Efron, "American Institute Science Clubs Exhibit at the World's Fair, 1939...Preliminary Report," [January 1939], NYHS AIR, Box 177, Folder 5, 1–5.

28. H. D. Lufkin to H. H. Sheldon, January 17, 1939, NYHS AIR, Box 376, Folder 13.

29. G. Edward Pendray to American Institute of the City of New York, January 27, 1939, NYHS AIR, Box 425, Folder 13; "Press Release," January 31, 1939, NYHS AIR, Box 425, Folder 13, 1–3.

30. The American Institute, "To the Board of Education of the City of New York A Proposal to Establish Student Science Activities at the World's Fair," NYHS AIR, Box 181, Folder 8, 2–3.

31. Institute members estimated the cost to the school board of hiring substitute teachers at $15,000. "The American Institute at the World's Fair," NYHS AIR, Box 181, Folder 8, 2–3.

32. "Junior Science Clubs at the World's Fair," *School and Society* 48 (July 2, 1938): 10. During his tenure, Campbell had also supported experimental programs in child-centered curricula and pedagogy in 70 of New York City's schools. See Diane Ravitch, *The Great School Wars: New York City, 1805–1973* (New York: Basic Books, 1974), 237.

33. Harold G. Campbell to the Principal and to the Heads of Science Departments, March, 9, 1939, NYHS AIR, Box 190, Folder 11.
34. "The American Institute World's Fair Project Dinner Meeting Chemist's Club," January 17, 1939, NYHS AIR, Box 214, Folder 30, 1–3.
35. "Press Release," January 30, 1939, NYHS AIR, Box 214, Folder 13; "Luncheon for Science Club Sponsors," January 28, 1939, NYHS AIR, Box 214, Folder 31, 5; Alexander Efron to H. H. Sheldon, February 18, 1939, NYHS AIR, Box 220, Folder 13, 2–4; "Visit the American Institute World's Fair Exhibit at the Westinghouse Building Arranged in Co-operation with the Board of Education of New York City," [1939], NYHS AIR, Box 180, Folder 6; "Club Members Ready for Demonstrations at N.Y. World's Fair," *Science Observer* 1 (April 1939): 1; Alexander Efron, "Science and Engineering at the World's Fair," March 1, 1939, NYHS AIR, Box 214, Folder 13, 1–2; "Science Exhibit of the American Institute," *School Science and Mathematics* 39 (March 1939): 225.
36. Publicity Department of the Westinghouse Electric and Manufacturing Company, "[Press Release]," 1939, JHHC GWM, Box 69, Folder, "Westinghouse Publicity," 2.
37. Publicity Department of the Westinghouse Electric and Manufacturing Company, "Westinghouse Ad Campaign Based on New York World's Fair," April 26, 1939, Westinghouse Electric Corporation Records, 1865–2000, MSS 424 housed at the Sen. John Heinz History Center, Pittsburgh, PA (hereafter JHHC WEC), Series VIII, Box 82, Folder 14, 1–3. The printed advertising series ran in popular magazines, including *Saturday Evening Post*, *Collier's*, and *Life*.
38. "Westinghouse at the World's Fair," *Westinghouse Magazine* 10 (November–December, 1938): 7–8. The quotation appears on page 7.
39. Publicity Department, Westinghouse Electric and Manufacturing Company, "Value of Electricity on Farm Is Demonstrated," [1939], JHHC WEC, Series IV, Box 70, Folder 6.
40. New York World's Fair 1939 Department of Feature Publicity, "Science at the New York World's Fair 1939," [1939], JHHC WEC, Series IV, Box 70, Folder 6, 32–33.
41. "Westinghouse at the World's Fair," [1938], JHHC WEC, Series IV, Box 70, Folder 8; Westinghouse Electric and Manufacturing Company, "Fair to Feature $50,000,000,000 'Electrorama,'" [February 1940], JHHC WEC, Series IV, Box 70, Folder 7, 1–3. Exhibits also aimed to bolster Westinghouse's ongoing campaign for rural electrification. See, for instance, "Rural Electrification Exhibit at the New York World's Fair," [1939], JHHC GWM, Box 69, Folder, "Rural Electrification."
42. "The Westinghouse Exhibit at the New York World's Fair 1939: A Hand-book for Employes [sic]," [1939], JHHC WEC, Series IV, Box 70, Folder 8, 22–41.

43. H. D. Lufkin to H. H. Sheldon, April 4, 1939, NYHS AIR, Box 367, Folder 13; Westinghouse Editorial Service, "High School Students Will Conduct Research at World's Fair," [1939], JHHC GWM, Box 69, Folder, "Junior Science Exhibit," 1–3.
44. Roland Marchand and Michael L. Smith, "Corporate Science on Display," in *Scientific Authority and Twentieth Century America*, ed. Ronald Walters, 148–184 (Baltimore: Johns Hopkins University Press, 1997); Steven Conn, *Museums and American Intellectual Life, 1876–1926* (Chicago: University of Chicago Press, 1998), 247; Steven Conn, *Do Museums Still Need Objects?* (Philadelphia, University of Pennsylvania Press, 2010), 161.
45. Warren I. Susman, "The People's Fair: Cultural Contradictions of a Consumer Society," in *Dawn of a New Day: The New York World's Fair, 1939/40*, ed. Helen A. Harrison, 16–27 (New York: New York University Press, 1980). See also, Larry Zim, *The World of Tomorrow: The 1939 New York World's Fair* (New York: Harper & Row, 1988); Joseph Philip Cusker, "The World of Tomorrow: The 1939 New York World's Fair," (PhD diss., Rutgers University, 1990); Carlos Emmons Cummings, *East Is East and West Is West: Some Observations on the World's Fairs of 1939 by One Whose Main Interest Is in Museums* (Buffalo: Buffalo Museum of Science, 1940); Kuznick, "Losing the World of Tomorrow," 341–373; Rydell, "The Fan Dance of Science," 525–542.
46. Eight out-of-town secondary schools either supplied exhibits or sent students to work at the science labs. These included, for example, a flower exhibit from David Prince Junior High School in Jacksonville, Illinois, a hot water system from the Elizabeth Peabody Settlement House in Boston, and a "mechanical smoker" from Eastern High School in Buffalo, New York. Cummings, *East Is East and West Is West*, 76; Hazel MacCallum, "World's Fair 1939 Project," NYHS AIR, Box 376, Folder 16; Marcia Roach to Hazel McCallum [sic], October 25, 1940, NYHS AIR, Box 214, Folder 9; Hazel MacCallum, "List of Exhibits from Clubs at a Distance at World's Fair 1939," NYHS AIR, Box 198, Folder 5; Charles Zavales, "Exhibits: List of Wall Exhibits in Junior Science Hall," October 3, 1939, NYHS AIR, Box 198, Folder 5, 1–3; "New Exhibits Added by Club Members at Junior Science Hall," *Science Observer* 1 (July–August 1939): 1, 3, 6, 8.
47. Kuznick, "Losing the World of Tomorrow," 341–373; Rydell, "The Fan Dance of Science," 536–539; Cummings, *East Is East and West Is West*, 74.
48. "Visitors to N.Y. Fair See Many Wonders Achieved by Science," *Science Observer* 1 (June 1939): 1, 8; Cusker, "The World of Tomorrow," 256; "Highlights and Shadows on the New York World's Fair," *Science Observer* 1 (May 1939): 5; "Science at the New York World's Fair 1939 Compiled by the Department of

Science and Education and the Department of Feature Publicity," 1939, JHHC GWM, Box 69, Folder "Fair Publicity," 45–46.
49. "Westinghouse Fair World: Official News of Westinghouse Activities at the New York and San Francisco Fairs," [1939], JHHC GWM, Box 69, Folder "Westinghouse Publicity"; Westinghouse Electric and Manufacturing Company Publicity Department, "Dishwashers Compete in Epic 'Battle of Centuries' at Fair," [1939], JHHC WEC, Series VIII, Box 82, Folder 14; Cummings, *East Is East and West Is West*, 110. Such presentations of science were emblematic of larger trends in the twentieth century. See John Burnham, *How Superstition Won and Science Lost: Popularizing Science and Health in the United States* (New Brunswick: Rutgers University Press, 1987); Bruce V. Lewenstein, "The Meaning of 'Public Understanding of Science' in the United States after World War II," *Public Understanding of Science* 1 (January 1992): 45–68.
50. "Junior Science Activities at New York World's Fair 1939," NYHS AIR, Box 214, Folder 12.
51. "The American Institute Science and Engineering Clubs' Exhibit at the World's Fair in Cooperation with the Board of Education of the City of New York Westinghouse Building," [1939], NYHS AIR, Box 209, Folder 9, 5.
52. "Youth's Fair within a Fair," *Science Observer* 1 (May 1939): iii.
53. Pierson's presentation had apparently impressed a wealthy spectator who secured a scholarship for him to attend a prestigious preparatory school in Pennsylvania. American Institute officials also arranged for Pierson to appear on the *Bright Ideas Club* radio program of the National Broadcasting Company. "14-year-old Chemist Wins Scholarship," *Science Observer* 1 (July–August 1939): 1, 8.
54. "Marconi Memorial Award—Saturday August 12—1:45 to 2:00 P. M. EDST," NYHS AIR, Box 409, Folder 6, 2–3; The American Institute of the City of New York, "Marconi Scholarship Award to be Given at Fair," August 11, 1939, NYHS AIR, Box 409, Folder 6; "American Institute Day at the Fair," *Science Observer* 1 (September 1939): 1, 7.
55. This represented a concern shared by World's Fair planners as a whole, because their projections well exceeded the actual attendance. Financially, the fair did not turn a profit in either 1939 or 1940.
56. G. Edward Pendray, "Report of the First Meeting of the Publicity Committee for the American Institute's Science and Engineering Club's Exhibit at the World's Fair," June 29, 1939, NYHS AIR, Box 213, Folder 8, 1–2; G. Edward Pendray, "Report of the Second Meeting of the Publicity Committee for the American Institute's Science and Engineering Club's Exhibit at the World's Fair," July 6, 1939, NYHS AIR, Box 213, Folder 8, 1–2; G. Edward Pendray, "Report and Recommendations of the Publicity Committee for the

American Institute's Science & Engineering Clubs' Exhibit at the World's Fair," July 13, 1939, NYHS AIR, Box 180, Folder 6, 1–4.
57. Howard Stephenson to G. Edward Pendray, September 25, 1939, JHHC GWM, Box 69, Folder "Westinghouse Publicity," 1–4; Howard Stephenson to G. Edward Pendray, August 18, 1939, JHHC WEC, Series IV, Box 70, Folder 11, 1–3; G. Edward Pendray to J. Gilbert Baird, July 28, 1939, JHHC WEC, Series IV, Box 70, Folder 11, 1–2.
58. Cusker, "The World of Tomorrow," 282–291.
59. H. H. Sheldon, "Form A," [1939], NYHS AIR Box 197, Folder 17; "Do You Have a Science Club?" *Science Observer* 1 (October 1939): 1; "Science Leaflet Will Serve Sponsors of the A. I. S. & E. Clubs," *Science Observer* 1 (October 1939): 1, 7; *112th Annual Report the American Institute of the City of New York for the Year Ending December 31, 1939* (New York, 1940), NYHS AIR, Box 376, Folder 16, 5–8.
60. The American Institute's Trustees' decision in 1939 to eliminate its very active associate membership program indicated the organization's precarious financial position. *112th Annual Report the American Institute*, 5–8; The American Institute of the City of New York, "The Eleventh Annual American Institute Science and Engineering Fair," [1939], SIA RU 7091, Box 205, Folder 6, 8–9; The American Institute of the City of New York, "The Twelfth Annual Science and Engineering Fair," [1940], NYHS AIR, Box 376, Folder 16, 1–16; "12th Annual Science and Engineering Fair," *Science Observer* 2 (May 1940): 8–10; "Membership in the A. I. S. & E. Clubs Was Trebled during Year," *Science Observer* 1 (November 1939): 1, 7.
61. Hazel MacCallum to H. H. Sheldon, "Suggestions for Junior Science Hall, World's Fair 1940," October 24, 1939, NYHS AIR, Box 197, Folder 17, 1–2. Emphasis in original.
62. Hazel MacCallum, "World's Fair 1940," January 12, 1940, NYHS AIR, Box 214, Folder 12, 1–2.
63. Westinghouse Electric and Manufacturing Company Publicity Department, "Television Thrills by Westinghouse for World's Fair Visitors," [1939], JHHC WEC, Series VIII, Box 82, Folder 14; Westinghouse Electric and Manufacturing Company Publicity Department, "Westinghouse at the New York World's Fair," [1939], JHHC WEC, Series VIII, Box 82, Folder 14, 1–8.
64. Frederic Ernst to Robert T. Pollock, February 27, 1940, NYHS AIR, Box 206, Folder 5, 1–2.
65. Ibid., 3–4.
66. H. D. Lufkin to H. H. Sheldon, February 29, 1940, NYHS AIR, Box 206, Folder 5.
67. Westinghouse Electric and Manufacturing Company, "Laboratory Marvels by Young Scientists to be Seen at Fair," [1939], JHHC GWM, Box 69, Folder "Junior Science Exhibit," 1–3.

68. John C. Burnham, *How Superstition Won and Science Lost*, 4–5, 170.
69. H. H. Sheldon to Frederic Ernst, March 14, 1940, NYHS AIR, Box 206, Folder 5, 1–3.
70. *113th Annual Report of the American Institute of the City of New York for the Year Ending December 31, 1940* (New York, 1941), NYHS AIR, Box 387, Folder 1, 5–7; "Coming Events Cast Their Shadows," *Science Observer* 2 (December 1940): 14–15; "What Tomorrow's Scientists Are Doing Today!" *Science Observer* 2 (June 1940): 3–6; "New York World's Fair 1940," NYHS AIR, Box 177, Folder 3, 1–4.
71. "Confidence in Americanism," *Science Observer* 2 (October 1940): 14–17; Henry Platt to H. D. Lufkin, June 27, 1940, NYHS AIR, Box 213, Folder 1.
72. "Radio Script Broadcast by Participants of the American Institute Student's Science Laboratory and Dr. H. C. Parmelee over WNYC, Saturday, August 3rd, 1940," NYHS AIR, Box 212, Folder 7, 6.
73. H. H. Sheldon, "The Science Club Program of the American Institute," *School Science and Mathematics* 40 (April 1940): 365–367. The quotation appears on pages 366–367.
74. G. Edward Pendray to Howard Stephenson, June 25, 1940, JHHC GWM, Box 69, Folder "Westinghouse Publicity," 2; "Youth Looks to the Future," September 23, 1940, NYHS AIR, Box 212, Folder 7, 1; Henry Platt to H. H. Sheldon, August 15, 1940, NYHS AIR, Box 212, Folder 7; "Building the Woman's World of Tomorrow," August 17, 1940, NYHS AIR, Box 212, Folder 7, 1–8.
75. "Youth Looks to the Future," 4–6.
76. "Confidence in Americanism," 14–15.
77. "U.S. Asks: What Is Your Hobby?" *Science Observer* 2 (November 1940): 14–15; "National Defense Becomes Our Task," *Westinghouse Magazine* 12 (August 1940): 1.
78. Richard Polenberg, *War and Society: The United States, 1941–1945* (Philadelphia: J.P. Lippincott, 1972); Geoffrey Perrett, *Days of Sadness, Years of Triumph: The American People, 1939–1945* (New York: Coward, McCann, & Geoghegan, 1973).
79. "Agenda for Meeting of the Committee of the Academy Conference and the American Institute," [1939], NYHS AIR, Box 413, Folder 21; E. C. L. Miller to H. H. Sheldon, January 18, 1940, NYHS AIR Box 413, Folder 21; "Pennsylvania Junior Academy Cooperates with A. I. S. & E. Clubs," *Science Observer* (September 1939): 1, 7; "Report of the High School Relations Committee—1939," *Proceedings of the Iowa Academy of Science for 1939* 46 (April 1939): 38–40; "Report of the High School Relations Committee 1939–1940," *Proceedings of the Iowa Academy of Science for 1940* 47 (April 1940): 27–29.
80. Bert Cunningham to H. H. Sheldon, March 20, 1940, NYHS AIR, Box 416, Folder 17, 1–2; Bert Cunningham to H. H. Sheldon,

August 23, 1940, NYHS AIR, Box 416, Folder 17, 1–6; Bert Cunningham to H. H. Sheldon, September 13, 1940, NYHS AIR, Box 416, Folder 17, 1–7; Bert Cunningham to H. H. Sheldon, March 13, 1941, NYHS AIR, Box 416, Folder 17, 1–2.

81. "Active Clubs Everywhere," *Science Observer* 1 (May 1939): 7; "Active Clubs Everywhere," *Science Observer* 1 (June 1939): 7; "Active Clubs Everywhere," *Science Observer* 2 (July–August 1939): 7; "Science Leaflet Will Serve Sponsors of the A. I. S. & E. Clubs," 1, 7.

82. "The Constructive Blitzkrieg: Science and Engineering Clubs Mushroom across the Country," *Science Observer* 2 (June 1940): 19. See also, "Confidence in Americanism," 14–17; "Membership in the A. I. S. & E. Clubs Was Trebled During Year," 1, 7.

83. *113th Annual Report of the American Institute*, 7; Joseph H. Kraus to H. H. Sheldon, July 2, 1940, NYHS AIR, Box 214, Folder 8, 1–2; Henry Platt, "A Few Suggestions for a Science Workshop and Center," July 5, 1940, NYHS AIR, Box 214, Folder 22, 1–6; Westinghouse Electric and Manufacturing Company, "Westinghouse to Aid 700 Science Clubs for U.S. Youth," December 6, 1940, NYHS AIR, Box 387, Folder 15, 1–4; American Institute of the City of New York, "Open Nation's First Science Workshop for Youth Today," NYHS AIR, Box 387, Folder 15, 1–4; "The American Institute's Own Laboratory," *Science Observer* 3 (March 1941): 18; Harrison Kinney, "The Year of the Gifted Children," *Think* 45 (September/October 1979): 12–17.

84. On Science Service's early organizational history and efforts at science popularization through commercial radio, see Marcel Chotkowski LaFollette, *Science on the Air: Popularizers and Personalities on Radio and Early Television* (Chicago: University of Chicago Press, 2008), especially chapters 3–5, 8–9. See also David J. Rhees, "A New Voice for Science: Science Service under Edwin E. Slosson, 1921–30," (MA thesis, University of North Carolina, 1979), 20–25, 61–70, 87–88; Marcel C. LaFollette, "Taking Science to the Marketplace: Examples of Science Service's Presentation of Chemistry during the 1930s," *HYLE—International Journal for Philosophy of Chemistry* 12 (June 2006): 67–97.

85. "Science Service Backs Science Clubs Movement," *Science News Letter* 40 (September 27, 1941): 204; Board of Trustees of the American Institute, "Meeting Minutes," January 21, 1942, SIA RU 7091, Box 231, Folder 6, 1–3; Board of Trustees of the American Institute, "Meeting Minutes," February 18, 1942, SIA RU 7091, Box 231, Folder 6, 1–3; Board of Trustees of the American Institute, "Meeting Minutes," April 15, 1942, SIA RU 7091, Box 231, Folder 6, 1–3.

Ironically, Davis had been a member of the American Institute and was awarded a fellowship in 1937 "for interpreting to the people of the Nation the rapid progress of science upon which modern civilization depends and for the organized dissemination of research findings as news." See "American Institute Awards to Bell Laboratories, Davis," *Science News Letter* 33 (January 9, 1937): 28.
86. Cusker, "The World of Tomorrow," 2.
87. David Kaiser, "Cold War Requisitions, Scientific Manpower, and the Production of American Physicists after World War II," *Historical Studies in the Physical and Biological Sciences* 33 (1) (2002): 131–159; John L. Rudolph, *Scientists in the Classroom: The Cold War Reconstruction of American Science Education* (New York: Palgrave, 2002); David M. Donahue, "Serving Students, Science, or Society? The Secondary School Physics Curriculum in the United States, 1930–65," *History of Education Quarterly* 33 (Fall 1993): 321–352; Leopold E. Klopfer and Audrey B. Champagne, "Ghosts of Crisis Past," *Science Education* 74 (April 1990): 133–154; Philip W. Jackson, "The Reform of Science Education: A Cautionary Tale," *Daedalus* 112 (Spring 1983): 143–166; Barbara Barksdale Clowse, *Brainpower for the Cold War: The Sputnik Crisis and National Defense Education Act of 1958* (Westport, CT: Greenwood Press, 1981). As an important exception, John L. Rudolph examines World War II precedents for subsequent reforms in scientific research and science education. See John L. Rudolph, "From World War to Woods Hole: The Use of Wartime Research Models for Curriculum Reform," *Teachers College Record* 104 (March 2002): 212–241.
88. The American Institute, "Plan for the Junior Science Center at the New York World's Fair 1939," NYHS AIR, Box 178, Folder 5; "Radio Script Broadcast by Participants of the American Institute Student's Science Laboratory and Dr. H. C. Parmelee over WNYC, Saturday, August 3rd, 1940," NYHS AIR, Box 212, Folder 7; "Junior Science Activities at New York World's Fair 1939," NYHS AIR, Box 209, Folder 9. On the views of professional scientists, see, for instance, Gerald Wendt, *Science for the World of Tomorrow* (New York: W.W. Norton & Company, 1939); Peter J. Kuznick, *Beyond the Laboratory: Scientists as Political Activists in 1930s America* (Chicago: University of Chicago Press, 1987); and Kuznick, "Losing the World of Tomorrow," 344–350.

4   ENLISTING SCIENCE EDUCATION FOR NATIONAL STRENGTH

1. American high schools had experienced a half-century of unprecedented popularity. Roughly seven percent of youth between 14

and 17 years of age enrolled in public high schools in 1890. By 1940, that figure exceeded 79 percent. The draft age lowered to 18 years in 1942, and the need for more factory workers shriveled the nation's unemployment rate from 10 percent in 1940 to 2 percent by 1944.

2. Gerard Giordano's study of American schools during World War Two II asserts that "conservative" coalitions of industrial, military, and other federal constituents reoriented public education to meet national security issues—at the expense of progressive pedagogy, curriculum, and participatory democracy. Charles Dorn's more recent examination argues that many elements of progressive education persisted during and beyond World War II despite the rise of military and national imperatives. Gerard Giordano, *Wartime Schools: How World War II Changed American Education* (New York: Peter Lang, 2004), 5–28, 149; Charles Dorn, *American Education, Democracy, and the Second World War* (New York: Palgrave Macmillan, 2007), 1–18. See also Ronald D. Cohen, "Schooling Uncle Sam's Children: Education in the USA, 1941–1945," in *Education and the Second World War: Studies in Schooling and Social Change*, ed. Roy Lowe, 46–58 (London and Washington, DC: Falmer Press, 1992); John W. Studebaker, "What the Secondary Schools Can Do to Help Win This War," *The Bulletin of the National Association of Secondary-School Principals* 26 (October 1942): 11–17; Stephen J. Wright, "Impact on the High-school Curriculum," *Journal of Educational Sociology* 16 (March 1943): 424–450; Stanford University School of Education Faculty, *Education in Wartime and After* (New York: D. Appleton-Century Company, 1943), 231–242.

3. For example, Davis delivered addresses about popularizing science and science education to a wide range of professional organizations, some of which included the American Academy of Arts and Sciences, the American Association for the Advancement of Science, the National Education Association (and National Science Teachers Association), and National Association of Science Writers.

4. Watson Davis, "The Popularization of Science," February 4, 1937, SIA RU 7091, Box 443, Folder 10, 1–2.

5. Watson Davis, "The Public's Way to Science," July 12, 1938, SIA RU 7091, Box 443, Folder 2.

6. Otis W. Caldwell to Watson Davis, April 28, 1934, SIA RU 7091, Box 152, Folder 12; Watson Davis, "Minutes Executive Committee of Science Service," August 7, 1935, SIA RU 7091, Box 3, Folder 8, 3; "Minutes of the Meeting of the Executive Committee of Science Service," October 26, 1935, SIA RU 7091, Box 3, Folder 8, 4; Watson Davis to Harlow Shapley, August 20, 1941, SIA RU 7091, Box 229, Folder 3; Watson Davis to Harlow

Shapley, August 21, 1941, SIA RU 7091, Box 229, Folder 3; Henry Platt to D. H. Drinkard, November 24, 1941, NYHS AIR, Box 224, Folder 4; Tom K. Phares, *Seeking—and Finding—Scientific Talent: A 50-Year History of the Westinghouse Science Talent Search* (Pittsburgh: Westinghouse Electric Corporation, 1990), 2–6; Joseph Berger, *The Young Scientists: America's Future and the Winning of the Westinghouse* (Reading, MA: Addison-Wesley, 1994), 17–18; Ron Cowen, "'Go for It, Kid.' Looking Back on Five Decades of the Science Talent Search," *Science News* 139 (February 23, 1991): 120–123; "Science Service Backs Science Clubs Movement," *Science News Letter* 40 (September 27, 1941): 204. On G. Edward Pendray, see Elizabeth Neuffer, "G. E. Pendray, 86, Rocket Proponent," *New York Times* (September 20, 1987): 60; "Biographical Material on G. Edward Pendray," 1945, The Westinghouse Educational Foundation Records Housed in the CBS Corporation Inc. Offices in Pittsburgh, PA (hereafter CBS WEF), Box "Ruch Docs, 2 of 2, Science Talent Search," Folder "G. Edward Pendray," 1–3.

7. "Westinghouse Science Talent Search," 1974, CBS WEF, Box "Ruch Docs, 2 of 2, Science Talent Search," Folder "Education—Science Talent Search—Scholarships," 1; Phares, *Seeking—and Finding— Scientific Talent*, 5–6; "Ten Scholarships Offered to Outstanding Seniors," *Science News Letter* 45 (February 5, 1944): 82; "Camp Fire Girl, 16, Wins Prize in Poetry Contest," *New York Times* (February 8, 1940): 15.
8. Watson Davis, "Executive Committee of Science Service," March 28, 1942, SIA RU 7091, Box 5, Folder 8, 2.
9. Watson Davis, "Report to the Annual Meeting of the Board of Trustees of Science Service, Thursday, April 30, 1942," 1942, SIA RU 7091, Box 5, Folder 8, 1–2.
10. Cowen, "'Go for It, Kid,'" 121.
11. David B. Tyack, *The One Best System: A History of American Urban Education* (Cambridge, MA: Harvard University Press, 1974), 182–183; John L. Rudolph, *Scientists in the Classroom: The Cold War Reconstruction of American Science Education* (New York: Palgrave, 2002), 9–81, 193–99; Kim Tolley, *The Science Education of American Girls: A Historical Perspective* (New York: RoutledgeFalmer, 2003), 196–206; David M. Donahue, "Serving Students, Science, or Society? The Secondary School Physics Curriculum in the United States, 1930–65," *History of Education Quarterly* 33 (1993): 321–352; Philip J. Pauly, "The Development of High School Biology: New York City, 1900–1925," *Isis* 82 (December 1991): 662–88. See also Victor H. Noll, *The Teaching of Science in Elementary and Secondary Schools* (New York: Longmans, Green, and Co., 1942); National Science Teachers Association. *Science Instruction for National Security: 1945 Yearbook*

(Washington, DC: National Science Teachers Association, 1945); Nelson B. Henry, *The Forty-sixth Yearbook of the National Society for the Study of Education: Part I Science Education in American Schools* (Chicago: University of Chicago Press, 1947).

12. Watson Davis, "War Service for Youth and Science: Science Clubs of America and the War Effort," December 30, 1941, SIA RU 7091, Box 445, Folder 13.
13. Joseph H. Kraus, "Amateur Scientists Can Help America: What Your Club Can Do to Serve," *Science News Letter* 41 (January 10, 1942): 24–25. The quotation appears on page 25.
14. Joseph H. Kraus, "Converting Scrap into Useful Articles," *Science News Letter* 41 (April 18, 1942): 248–249; Joseph H. Kraus, "Conserve by Learning to Splice Rope," *Science News Letter* 41 (May 16, 1942): 312–313; Joseph H. Kraus, "First Aid in War Emergency," *Science News Letter* 41 (March 14, 1942): 168–169; Joseph H. Kraus, "Why an Acoustic Mine Explodes," *Science News Letter* 40 (November 8, 1941): 296–297; Joseph H. Kraus, "Training for Youth and War Workers," *Science News Letter* 42 (November 21, 1942): 328–329.
15. "Science Clubs of America to Cooperate with Navy," *Science News Letter* 41 (February 14, 1942): 104–105; "News of Clubs," *Science News Letter* 41 (February 7, 1942): 92; J. K. Cannon to Science Clubs of America, May 11, 1942, RU 7091, Box 237, Folder 7; Ralph T. Millet to Joseph Kraus, August 12, 1942, SIA RU 7091, Box 237, Folder 7.
16. "News of Clubs," *Science News Letter* 41 (April 4, 1942): 221; "News of Clubs," *Science News Letter* 41 (January 17, 1942): 47; "Science Clubs of America," *Science News Letter* 41 (January 31, 1942): 78; "News of Clubs," *Science News Letter* 41 (June 20, 1942): 397; "News of Clubs," *Science News Letter* 41 (March 7, 1942): 159; "News of Clubs," *Science News Letter* 41 (January 17, 1942): 47.
17. "News of Clubs," *Science News Letter* 41 (April 11, 1942): 238.
18. "News of Clubs," *Science News Letter* 41 (May 30, 1942): 351.
19. "Science Clubs Help in War," *Science News Letter* 43 (April 3, 1943): 220; "News of Clubs," *Science News Letter* 43 (March 28, 1942): 206.
20. Watson Davis, "War—It's Technical," *Science News Letter* 42 (November 21, 1942): 329.
21. *Adventures in Science*, September 18, 1943, SIA RU 7091, Box 390, Folder 21, 4.
22. "What 4,400 Science Clubs Do," *Science News Letter* 45 (May 27, 1944): 350.
23. See, for example, "News of Clubs," *Science News Letter* 40 (December 13, 1941): 383; "News of Clubs," *Science News Letter* 40 (December 27, 1941): 411; "News of Clubs," *Science News Letter* 41 (January 3, 1942): 13; "News of Clubs," *Science News Letter* 41

(February 28, 1942): 143; "News of Clubs," *Science News Letter* 41 (May 9, 1942): 300; "News of Clubs," *Science News Letter* 41 (May 23, 1942): 333; "News of Clubs," *Science News Letter* 41 (June 13, 1942): 383.
24. "Science Clubs Help in War," *Science News Letter* 43 (April 3, 1943): 220; "What 4,400 Science Clubs Do," 350; Science Clubs of America, "You and Your Fellow Teachers Can Rightly Be Proud of the Results You Are Getting—and Can Look Forward to the Thanks of Grateful Students," 1946, SIA RU 7091, Box 446, Folder 17.
25. Science Clubs of America, "SCA Co-projects," 1945, SIA RU 7091, Box 394, Folder 19, 2. Some co-projects did not link to the war effort. For example, J. Edgar Hoover, director of the Federal Bureau of Investigation, solicited clubs' reports on their methods of fingerprinting, handling of evidence, laboratory examinations of blood and firearms, and the applications of physical science in crime detection.
26. Science Clubs of America, "SCA Co-projects," 5.
27. Mark H. Leff, "The Politics of Sacrifice on the American Home Front in World War Two," *Journal of American History* 77 (March 1991): 1296–1318.
28. Science Clubs of America, "SCA Co-projects," 9.
29. "Shortage of Physicists a National Emergency," *Science News Letter* 42 (August 1, 1942): 78.
30. James P. Mitchell, "Pre-induction Training," *Science News Letter* 42 (November 14, 1942): 314–315. See also "Science from Shipboard," *Science News Letter* 43 (May 1, 1943): 278–279.
31. "25,000 Women Will Be Employed by Armed Forces," *Science News Letter* 41 (March 28, 1942): 196; "Woman Engineers Needed," *Science News Letter* 42 (December 26, 1942): 407.
32. *Adventures in Science*, April 17, 1943. The Papers of Harlow Shapley, Harvard University Archives, Cambridge, MA [hereafter HUA], HUG 4773.10, Box 4E, Folder, "Radio Talks Folder #2."
33. "News of Clubs," *Science News Letter* 41 (March 21, 1942): 190.
34. "News of Clubs," *Science News Letter* 41 (May 2, 1942): 286.
35. "Young Scientists Work," *Science News Letter* 44 (October 16, 1943): 254–255.
36. "Prospecting for Future Scientists," *Science News Letter* 43 (April 10, 1943): 235.
37. Westinghouse Educational Foundation, "Scholarships by Westinghouse," [c. 1952], CBS WEF, Box "Ruch Docs, 2 of 2, Science Talent Search," Folder, "Education—Scholarships," 1–12; "Company Begins to Train Girl Engineering Assistants," *Westinghouse Magazine* 15 (September 1943): 10.
38. The Westinghouse Electric Corporation constituted the sole donor to the foundation, and it furnished an initial sum of $1,070,000.

"Proposal for the Establishment of a Westinghouse Educational Foundation," [1944], CBS WEF, Box "Jon Olsen's Office," Folder "WEDND—Deed w/ Trust," 1; "Westinghouse Educational Foundation," [c. 1944], CBS WEF, Box "Jon Olsen's Office," Folder "WEDND—Deed w/ Trust."

39. Educational Department Westinghouse Electric Corporation, "Scholarships by Westinghouse," [1944], CBS WEF, Box "Ruch Docs, 2 of 2, Science Talent Search," Folder "Education—Scholarships"; "Scholarships by Westinghouse," [c. 1946], CBS WEF, Box "Ruch Docs, 2 of 2, Science Talent Search," Folder "Education—Scholarships," 1–13; "First George Westinghouse Scholars Graduate," *Westinghouse Magazine* 15 (January 1943): 14, 30; "Proposal for the Establishment of a Westinghouse Educational Foundation," 3.

40. Davis, for instance, asked Pendray for the names of Westinghouse executives to receive copies of *Science News Letter*. G. Edward Pendray to Watson Davis, October 1, 1942, SIA RU 7091, Box 224, Folder 3, 1–2.

41. "Reykjavik Off the Port Bow!" *Science News Letter* 41 (February 14, 1942): 107. For similar themes, see "Science Shoulders Arms," *Science News Letter* 41 (January 10, 1942): 27; "Throats to Speak Our Nation's Piece," *Science News Letter* 41 (April 11, 1942): 237; "This Is the Way to Win a Battle in the Desert," *Science News Letter* 43 (March 13, 1943): 171; "Bombers from the Bottom of the Deep Blue Sea," *Science News Letter* 42 (July 11, 1942): 29.

42. "Bad Medicine for Big Bombers," *Science News Letter* 41 (March 14, 1942): 173.

43. "On Guard against Electrical Blackouts," *Science News Letter* 42 (September 12, 1942): 176. See also "We Think This Is What They Meant," *Science News Letter* 41 (May 9, 1942): 303; "Abandon Hope...All Germs Who Enter Here," *Science News Letter* 42 (October 10, 1942): 235; "Dust Takes a Holiday," *Science News Letter* 42 (August 8, 1942): 91; "Keeping Fugitive Carbon on the Job," *Science News Letter* 43 (February 13, 1943): 107; "What's a Cotangent Got to Do with Ack-ack?" *Science News Letter* 43 (May 8, 1943): 301; "Speaking of Superior Races...," *Science News Letter* 43 (June 12, 1943): 379; "What Goes on under a Nazi Pilot's Cap?" *Science News Letter* 44 (August 7, 1943): 91; "U.S. Tank Is Best, Say Nazi Experts," *Science News Letter* 44 (September 18, 1943): 188; "Periscope on the Starboard Quarter!" *Science News Letter* 44 (October 10, 1943): 27; "The Rubber Plant with Roots Two Miles Deep!" *Science News Letter* 44 (November 13, 1943): 315.

44. "Wanted: Future Faradays and Curies," *Science News Letter* 41 (June 13, 1942): 381.

45. "40 High School Seniors on the Way Up," *Science News Letter* 42 (November 7, 1942): 304. See also "The Second Annual Science

Talent Search Conducted by Science Clubs of America for the Westinghouse Science Scholarships," *Science News Letter* 42 (October 31, 1942): 280–281.
46. "Prospecting for Future scientists," 235; "For Forty High School Seniors Every Year, the Science Talent Search Leads to Washington!" *Science News Letter* 44 (October 16, 1943): 251; "A Nation-wide Search for Scientists of Tomorrow," *Science News Letter* 48 (August 18, 1945): 98.
47. "Westinghouse Research Accepts Every Wartime Challenge," *Science News Letter* 44 (December 11, 1943): 2.
48. Watson Davis, "Living in a Scientific World," February 28, 1942, SIA RU 7091, Box 445, Folder 7, 4–5. See also Science Service, "Use of Scientific Brain Power in the War," 1942, SIA RU 7091, Box 242, Folder 5, 1–8.
49. Harold A. Edgerton and Steuart Henderson Britt, "The First Annual Science Talent Search," *American Scientist* 31 (January 1943): 55–68. The quotation appears on page 55.
50. Harold A. Edgerton and Steuart Henderson Britt, "How Science Talent Search Winners Are Chosen," in *Science and the Future: Essays of the Winners of the Westinghouse Science Scholarships in the Second Annual Science Talent Search*, ed. Science Clubs of America (Washington, DC: Science Service, 1943), 112.
51. Edgerton and Britt, "The First Annual Science Talent Search," 55–62; Harold A. Edgerton and Steuart Henderson Britt, "The Science Talent Search," *Occupations* 22 (December 1943): 177–180; Edgerton and Britt, "How Science Talent Winners Are Chosen," 112–115.
52. *Adventures in Science*, July 4, 1942, SIA RU 7091, Box 389, Folder 26; "10,000 High School Seniors in Science Talent Search," *Science News Letter* 41 (May 30, 1942): 343.
53. *Adventures in Science*, July 4, 1942.
54. Watson Davis, "Foreword" in Science Service, *Youth Looks at Science and War: A collection of essays by the Washington Trip Winners of the First Annual Science Talent Search Conducted by Science Clubs of America* (Washington: Science Service and Penguin Books, 1942), vi. This volume circulated widely. A total of 14,000 copies were printed, nearly 5,000 of which were distributed to libraries established for soldiers at home and abroad. Watson Davis to Harlow Shapley, July 7, 1942, SIA RU 7091, Box 242, Folder 6.
55. Watson Davis, "Progress in the Sciences," 1942, SIA RU 7091, Box 445, Folder 18, 8.
56. Davis, "Foreword," iv; Science Service, "Information Memorandum on Progress of Science Service," October 24, 1942, SIA RU 7091, Box 5, Folder 7, 1–2.
57. Science Clubs of America, *Science and the Future: Essays of the Winners of the Westinghouse Science Scholarships in the Second Annual Science*

Talent Search (Washington, DC: Science Clubs of America, 1943), 126. A total of 48,000 copies of this volume were printed in 1943.
58. Ibid., vi.
59. "Winners in Research," *Science News Letter* 44 (October 23, 1943): 266.
60. "Science Talent Institute," *Science News Letter* 43 (March 6, 1943): 148.
61. Ibid.
62. M. L. Wilson, "Food Is a War Weapon," *Science News Letter* 43 (March 6, 1943): 150.
63. Karl T. Compton, "Young Scientists and War," *Science News Letter* 45 (March 18, 1944): 180. See also R. G. Robinson, "Aeronautical Engineering Scientific Methods," *Science News Letter* 45 (March 18, 1944): 182–183.
64. *Adventures in Science*, March 3, 1945, SIA RU 7091, Box 391, Folder 3.
65. Ibid.
66. "Science Talent Institute," *Science News Letter* 45 (March 11, 1944): 166–167. The quotation appears on page 167.
67. Thomas Parran, "Science and the Future," *Science News Letter* 43 (March 13, 1943): 172–174. Emphasis in the original.
68. Margaret Mead, "Women Can Help Apply Science to Human Problems," *Science News Letter* 45 (March 18, 1944): 189–190. The quotation appears on page 189. Addresses by a host of other scientists from 1943 and 1944 articulated similar themes. See W. C. Lowdermilk, "Americans Should Attack Soil Erosion Causes," *Science News Letter* 45 (March 18, 1944): 183; addresses by Henry C. Sherman, Bart J. Bok, and Marshall H. Stone appear in "Science Talent Institute," *Science News Letter* 45 (March 11, 1944): 166–167; S. G. Hibben, "Better Lighting Promised for World of Future," *Science News Letter* (March 18, 1944): 182; and Edwin G. Conklin, "The Biological Future," *Science News Letter* 43 (March 6, 1943): 149–150.
69. Ruth Hulda Miles, "Women in Medical Research," in Science Clubs of America, *Scientists of Tomorrow: Essays of the Winners of the Westinghouse Science Scholarships in the Third Annual Science Talent Search* (Washington, DC: Science Service, 1944), 74–76. See also "Science Calls Youth," *Science News Letter* 46 (October 7, 1944): 230–231.
70. Watson Davis, "Foreword," in Science Clubs of America, *Scientists of Tomorrow* (1944), iii.
71. Watson Davis, "Foreword," in Science Clubs of America, *Scientists of Tomorrow: Essays of the Winners of the Westinghouse Science Scholarships in the Fourth Annual Science Talent Search* (Washington, DC: Science Service, 1945), iv.

72. Watson Davis, "Science and the Press," *The Annals of the American Academy of Political and Social Science* 219 (January 1942): 100–106. The quotation appears on page 100.
73. *Adventures in Science*, March 28, 1942, SIA RU 7091, Box 389, Folder 12, 4. See also "Unified Science Must Serve Unified World after War," *Science News Letter* 41 (April 4, 1942): 214.
74. Watson Davis, "Science Clubs of America – An Educational Force for the Future," March 1945, SIA RU 7091, Box 442, Folder 22, 2.
75. Watson Davis, "Science Teaching and Science Clubs Now and Postwar," *School Science and Mathematics* 14 (March 1945): 257–264.
76. James Bryant Conant, "Science and Society in the Post-War World," *Vital Speeches of the Day* 9 (April 15, 1943): 394–397.
77. Franklin D. Roosevelt to Vannevar Bush, November 17, 1944, HUA HUG 4773.10, Box 17D, Folder "Bush-Moe-Committee."
78. Accounts of the Bush Report can be found in Daniel J. Kevles, "The National Science Foundation and the Debate over Postwar Research Policy, 1942–1945: A Political Interpretation of *Science— The Endless Frontier*," *Isis* 68 (March 1977): 5–26; Rudolph, *Scientists in the Classroom*, 41; and Joel Spring, *The Sorting Machine: National Educational Policy Since 1945* (New York: David McKay Company, 1976), 74–77.
79. Vannevar Bush to Harlow Shapley, December 14, 1944, HUA HUG 47730.10, Box 17D, Folder "Bush-Moe-Committee," 1–2.
80. Harlow Shapley, "Memorandum for the Bush Committee: Notes on Publicity in the Interests of Discovering and Encouraging Scientific Talent," 1944, HUA HUG 4773.10, Box 1B, Folder "Bush Committee. Notes on Publicity in the Interests of Discovering and Encouraging Scientific Talent," 1–5. The quotation appears on page 3.
81. Harlow Shapley, "Notes on Publicity in the Interests of Discovering and Encouraging Scientific Talent," March 1, 1945, HUA HUG 4773.10, Box 17D, Folder "Bush-Moe-Committee," 1–4. The quotation appears on pages 3–4.
82. Harlow Shapley, "Notes on the Program for Discovering and Developing Scientific Talent in American Youth in Order that We May Have Sufficient Trained Scientific Manpower for the Scientific Needs of the Future ... F. D. R.," December 17, 1944, HUA HUG 4773.10, Box 17D, Folder "Bush-Moe-Committee," 1–2.
83. Watson Davis, "Role of Science Clubs in the Development of Science Talent," January 15, 1945, HUA HUG 4773.10, Box 17D, Folder "Bush-Moe-Committee," 1.
84. Ibid., 2–3.
85. Ibid., 2.

86. James B. Conant to Henry Allen Moe, December 21, 1944, HUA HUG 4773.10, Box 17D, Folder "Bush-Moe-Committee," 1.
87. Ibid., 2. Conant's concerns about admissions criteria at Harvard University are elaborated in Jerome Karabel, *The Chosen: The Hidden History of Admission and Exclusion at Harvard, Yale, and Princeton* (New York: Houghton Mifflin, 2005), 166–169. Conant would continue to lament the unequal educational opportunities for American high school students in the coming years. See Wayne J. Urban, *More Than Science and Sputnik: The National Defense Education Act of 1958* (Tuscaloosa: University of Alabama Press, 2010), 83–84.
88. George K. Bennet, John M. Stalnaker, and Dale Wolfle, "Program for the Selection of Scientific Talent," [c. 1945], HUA HUG 4773.10, Box 17D, Folder "Bush-Moe-Committee," 1–2; Henry Allen Moe, "Henry Allen Moe to the Committee," April 27, 1945, HUA HUG 4773.10, Box 17D, Folder "Bush-Moe-Committee," 1–2.
89. Harlow Shapley to Henry Allen Moe, June 5, 1945, HUA HUG 4773.10, Box 17D, Folder "Bush-Moe-Committee," 1; Henry Allen Moe, "Partial First Draft of a Report to Dr. Bush," February 27, 1945, HUA HUG 4773.10, Box 17D, Folder "Bush-Moe-Committee," 1–20.
90. Vannevar Bush, "V. Bush to the Chairman and Members, Committees on Reply to the President's Letter," April 14, 1945, HUA HUG 4773. 10, Box 17D, Folder "Bush-Moe-Committee"; Vannevar Bush to Harlow Shapley, July 14, 1945, HUA HUG 4773.10, Box 17D, Folder "Bush-Moe-Committee."
91. H. A. Moe, "Report of the Committee on Discovery and Development of Scientific Talent," in Vannevar Bush, *Science – The Endless Frontier*, ed. Vannevar Bush, 136–185 (Washington, DC: U.S. Government Printing Office, 1945). The quotations are from pages 137, 139, 147, and 140.
92. Ibid., 137, 143, 149–150.
93. Ibid., 166–176, 181–182.
94. Ibid., 152–157, 183.
95. Vannevar Bush, "Research Agency Planned," *Science News Letter* 48 (July 28, 1945): 51–53.
96. Watson Davis, "Critical Shortage," *Science News Letter* 48 (August 25, 1945): 122–124.
97. Ibid., 123.
98. Ibid., 124. Davis also called on the National Science Teachers Association (NSTA) to support the recommendations in Bush's report. See NSTA, "Toward Building America," [1945], SIA RU 7091, Box 445, Folder 35; Watson Davis, "Federal Support for Postwar Scientific Research and Education," *The Educational Record* 26 (October 1945): 281–287.
99. David M. Kennedy, *Freedom from Fear: The American People in Depression and War, 1929–1945* (New York: Oxford University

Press, 2005), 747; Dorn, *American Education, Democracy, and the Second World War*, 7–8; Alan Brinkley, "World War II and American Liberalism," in *The War in American Culture: Society and Consciousness during World War II*, ed. Lewis A. Erenberg and Susan E. Hirsch, 313–330 (Chicago: University of Chicago Press, 1996).
100. Dorn, *American Education, Democracy, and the Second World War*, 16–17, 176. The quotation appears on page 176. Historian Ronald Cohen has similarly argued that "the War had touched the nation's schools in diverse ways, without leaving much of a visible imprint, or perhaps even legacy." Cohen, "Schooling Uncle Sam's Children," 55.
101. Giordano, *Wartime Schools*, xix–xxiii, 239–241. Steven Mintz has emphasized that for most American youth, the war comprised "the formative experience of their lives." Steven Mintz, *Huck's Raft: A History of American Childhood* (Cambridge, MA: Harvard University Press, 2004), 255.

5 Sustaining Mobilization in an Atomic Age

1. Frank Thone, "New Responsibilities," *Science News Letter* 48 (August 18, 1945): 100.
2. Watson Davis, "Science Previews," *Science News Letter* 49 (January 5, 1946): 10.
3. Paul Boyer, *By the Bomb's Early Light: American Thought and Culture at the Dawn of the Atomic Age* (Chapel Hill: University of North Carolina Press, 1994), 49–65.
4. Jessica Wang, *American Science in an Age of Anxiety: Scientists, Anticommunism, & the Cold War* (Chapel Hill: University of North Carolina Press, 1999); James T. Patterson, *Grand Expectations: The United States, 1945–1974* (New York: Oxford University Press, 1996), 128–178; Ellen Schrecker, *Many Are the Crimes: McCarthyism in America* (Princeton: Princeton University Press, 1998), 154–183; Boyer, *By the Bomb's Early Light*, 84–103, 335–339; David M. Hart, *Forged Consensus: Science, Technology, and Economic Policy in the United States, 1921–1953* (Princeton: Princeton University Press, 1998), 176–233.
5. Watson Davis, "Tomorrow's Scientists," November 29, 1946, SIA RU 7091, Box 446, Folder 14, 2.
6. Watson Davis, "Science and Libraries Today," March 20, 1947, SIA RU 7091, Box 446, Folder 17, 1.
7. "Science Review for 1946," *Science News Letter* 50 (December 21, 1946): 389–396; Watson Davis, "Science Previews for 1947," *Science News Letter* 51 (January 4, 1947): 10–12.
8. "Soviet Science Celebration," *Science News Letter* 47 (June 9, 1945): 358.

9. Harlow Shapley, "Science and Internationalism" radio script for *Adventures in Science*, June 20, 1945, HUA HUG 4773.10, Box 4E, Folder "Radio Talks Folder #2," 3–7.
10. *Adventures in Science*, April 20, 1946, RU 7091, Box 392, Folder 11, 2. See also "International Studies," *Science News Letter* 49 (April 27, 1946): 268; "H. Res 215," April 9, 1945, HUA HUG 4773.10, Box 4E, Folder "Radio Talks Folder #2," 1–2; Joann Palmeri, "An Astronomer beyond the Observatory: Harlow Shapley as Prophet of Science" (PhD diss., University of Oklahoma, 2000), 86–131.
11. Harlow Shapley, "The Scientist Outside the Laboratory," *The American Scholar* 15 (Autumn 1946): 411–415; Harlow Shapley to Watson Davis, April 22, 1947, HUA HUG 4773.10, Box 15D, Folder "Science Service—Watson Davis," 1.
12. Harlow Shapley, "The Cold War," [c.1949], HUA HUG 4773.10, Box 1B, Folder "Cold War," 1.
13. Shapley was the subject of Congressional investigations for continuing to advocate for international collaborations in science. Wang, *American Science in an Age of Anxiety*, 118–147; Schrecker, *Many Are the Crimes*, 164–183; Patterson, *Grand Expectations*, 204–205.
14. Margaret Patterson and Frank Thone, "Scientists of the Future," *Science News Letter* 48 (September 22, 1945): 186–188. The quotation appears on page 186.
15. Ibid., 187–188. See also Margaret E. Patterson, "Shortage of Scientists," *Science News Letter* 50 (October 5, 1946): 218–220.
16. "Scientists Deferred," *Science News Letter* 48 (November 17, 1945): 311. Shapley, meanwhile, called for federally issued draft deferments to promising scientists, but not expressly for bolstering national defense. Harlow Shapley, "Status Quo or Pioneer? The Fate of American Science," *Harper's Magazine* 191 (September 1945): 312–317.
17. Watson Davis, "Senate Votes Foundation," *Science News Letter* 50 (July 13, 1946): 19, 23.
18. U.S. Senate Subcommittee on War Mobilization, *Hearings on Science Legislation: S. 1297 and Related Bills* (Washington, DC: United States Government Printing Office, 1946), 1197.
19. Ibid., 1198, 1201, 1204–1207. Davis summoned other talent search winners to support this agenda. On the March 30, 1946, radio broadcast of *Adventures in Science*, for example, Jules A. Kernen from Southwest High School in St. Louis proposed that the pending National Science Foundation would "give training in college, to not just a few, but to several thousands of students each year." *Adventures in Science*, March 30, 1946, SIA RU 7091, Box 392, Folder 8, 6. See also *Adventures in Science*, February 19, 1949, SIA RU 7091, Box 396, Folder 30. On the political negotiations over the National Science Foundation, see Jessica Wang,

"Liberals, the Progressive Left, and the Political Economy of Postwar American Science: The National Science Foundation Debate Revisited," *Historical Studies in the Physical and Biological Sciences* 26 (1) (1995): 139–166; Daniel J. Kevles, "The National Science Foundation and the Debate over Postwar Research Policy, 1942–1945: A Political Interpretation of *Science—The Endless Frontier*," *Isis* 68 (March 1977): 4–26.
20. Watson Davis, "Cancers of Scientific Ignorance," June 13, 1946, SIA RU 7091, Box 446, Folder 21, 4–5.
21. National Science Teachers Association, *Science Instruction for National Security* (Washington, DC: NSTA, 1945), 1.
22. Ibid. The caption for the cover page appears on page 48.
23. Ibid., 6.
24. NSTA, *Time for Science Instruction* (Washington, DC: NSTA, 1946), 5.
25. The Bronx High School of Science admitted only one in three applicants. Morris Meister, "We Have Lost a Generation," *The Science Teacher* 14 (April 1947): 62; Morris Meister, "Science in Elementary and Secondary Education," *The Science Teacher* 15 (February 1948): 13–16; National Science Teachers Association, *Science Education for National Security*, 12–17.
26. Morris Meister, "Teaching Talented Youth in a Free Society," in *Science Education for National Security*, National Science Teachers Association, 13.
27. Morris Meister, "The Problem," in *Time for Science Instruction*, NSTA, 4 (Washington, DC: NSTA, 1946).
28. Morris Meister, "Outlook in Science Education," *The Science Teacher* 15 (October 1948): 105–107, 132.
29. Meister, "Teaching Talented Youth in a Free Society," 13. See also Meister, "Science in Elementary and Secondary Education," 15; National Science Teachers Association, *Science Instruction for National Security*, 27–32.
30. Science Clubs of America, "You and Your Fellow Teachers Can Rightly Be Proud of the Results You Are Getting—and Can Look Forward to the Thanks of Grateful Students," 1946, SIA RU 7091, Box 446, Folder 17; Science Clubs of America, "Do Boys and Girls Appreciate the Help They Get from Their Science Teachers?" [1946], SIA RU 7091, Box 446, Folder 17; "Science Club No. 10,000," *Science News Letter* 50 (July 20, 1946): 36; Watson Davis, "The World Is Unfinished," 1946, SIA RU 7091, Box 446, Folder 11, 2; "Science Clubs Abroad," *Science News Letter* 56 (August 6, 1949): 85.
31. Watson Davis, "Our Most Precious Resource," November 20, 1946, SIA RU 7091, Box 446, Folder 3, 3.
32. Harlow Shapley, "WCOP Radio Talk," June 8, 1946, HUA HUG 4773.10, Box 4E, Folder "Radio Talks Folder #1," 1; Margaret E.

Patterson, "Science Clubs Expand Their Activities," *The Science Teacher* 15 (October 1949): 129–131.
33. Harlow Shapley, "The Uses and Hopes of Scientific Societies," [c. 1947], HUA HUG 4773.75, Box "Manuscripts and Transcripts," Folder "Manuscripts and Transcripts," 1–7.
34. Watson Davis, "Searching for Science Talent," [c. 1945], SIA RU 7091, Box 44, Folder 35, 6.
35. Davis, "The World Is Unfinished," 2–3.
36. Watson Davis and Margaret Patterson, "Finding Young Scientists," [c. 1947], SIA RU 7091, Box 446, Folder 20, 1–4.
37. Margaret E. Patterson, "Clubwork Makes Science Fun," *Science News Letter* 56 (September 17, 1949): 186.
38. John L. Rudolph, *Scientists in the Classroom: The Cold War Reconstruction of American Science Education* (New York: Palgrave Press, 2002), 124, 169–170; *Adventures in Science*, September 23, 1950, SIA RU 7091, Box 397, Folder 15, 7–8; Watson Davis, "To Interest People in the Sciences," 1951, HUG 4773.10, Box 15D, Folder "Annual Report of the Board of Trustees of Science Service, Inc.," 2–3; Watson Davis, "For Science Understanding," *Science News Letter* 60 (July 21, 1951): 37–38.
39. Shapley was instrumental in creating UNESCO, and he referred to the proposed activities of the fledgling organization as contributing "toward the One-World idea." Shapley, "The Uses and Hopes of Scientific Societies," 5–7.
40. "Science Clubs of America to Cooperate with UNESCO," *Science News Letter* 54 (November 6, 1948): 296.
41. "Science Clubs Abroad," 85.
42. By the early 1950s, Davis indicated Science Service's opposition to a rivaling global organization of science clubs that would capitalize on the precedents established by SCA. Patterson, "Science Clubs Expand Their Activities," 129; United Nations Educational, Scientific and Cultural Organization, "International Meeting of Science Club Leaders," July 16, 1949 (afternoon session), SIA RU 7091, Box 418, Folder 14, 1–5; Watson Davis to Kees Meijers, July 31, 1951, SIA RU 7091, Box 305, Folder 1, 1–3; Watson Davis, "Report to the Annual Meeting of the Board of Trustees of Science Service, Sunday, April 27, 1952," 1952, HUA HUG 4773.10, Box 15D, Folder "Annual Report of the Board of Trustees of Science Service, Inc.," 5.
43. Margaret Patterson, "Modern Science and Science Clubs," October 26, 1945, SIA RU 7091, Box 446, Folder 3, 2.
44. Watson Davis, "The Frontiers of Science Are Still Endless," June 25, 1946, SIA RU 7091, Box 446, Folder 12, 2. See also Science Clubs of America, *How you can SEARCH for Science Talent: A Book of Facts about the Sixth Annual Science Talent Search for Westinghouse Scholarships* (Washington, DC: Science Service, 1946), 2.

45. Davis, "Our Most Precious Resource," 1.
46. Science Clubs of America, *How you can SEARCH for Science Talent* (1946), 4.
47. Watson Davis, "Youth Learns Science," *Science News Letter* 52 (October 4, 1947): 218–219. The quotation appears on page 219.
48. The intellectual engagement and vocational guidance of especially talented students in high school science clubs were therefore essential. Davis and Patterson, "Finding Young Scientists," 4–5; Davis, "The World Is Unfinished," 3; Watson Davis, "Summary of Comments," 1947, SIA RU 7091, Box 446, Folder 17, 1–4.
49. Wang, *American Science in an Age of Anxiety*, 250–293; Jessica Wang, "Scientists and the Problem of the Public in Cold War America, 1945–1960," *Osiris* 17 (2002): 323–347; Wang, "Liberals, the Progressive Left, and the Political Economy of Postwar American Science," 139–166. On the rationale behind civil defense education, see JoAnne Brown, "'A Is for Atom, B Is for Bomb': Civil Defense in American Public Education, 1948–1963," *Journal of American History* 75 (June 1988): 68–90.
50. Margaret E. Patterson, "Awards Await Winners," *Science News Letter* 56 (October 15, 1949): 250–251. The quotation is from page 251. See also Watson Davis, "Report to the Annual Meeting of the Board of Trustees of Science Service, Sunday, April 26, 1953," 1953, HUA HUG 4773.10, Box 15D, Folder "Science Service, Inc.—Annual Meeting of the Board of Trustees [4/26/53]," 1–7.
51. Davis sought $100,000 annually to expand club activities and services and believed a minimum of $25,000 was required to sustain them. Science Service had not fared well financially in the aftermath of World War II and endured a loss of net worth and income of $30,171.52 between 1947 and 1953. Davis, "To Interest People in the Sciences," 5; "Financial Progress of Science Service, Inc.," [1953], HUA HUG 4773.10, Box 15D, Folder "Science Service—[1951–1955]."
52. Watson Davis, "Report Upon $10,000 Grant from National Science Foundation to Science Service, Inc., for Support of Science Clubs of America," April 22, 1953, HUA HUG 4773.10, Box 15D, Folder "Science Service, Inc.—Annual Meeting of the Board of Trustees [4/26/53]," 1–4; "Aiding Young Scientists," *Science News Letter* 54 (October 16, 1948): 245–246; "At No Cost Whatever You Can Join Science Clubs of America the Largest Scientific Organization in the World," *Science News Letter* 60 (October 27, 1951): 269; Margaret E. Patterson and Joseph H. Kraus, *Thousands of Science Projects* (Washington, DC: Science Service, 1956), 45; Science Clubs of America, *Sponsor Handbook for 1958* (Washington, DC: Science Service, 1957), 1–66; Gilbert Benowitz, "Science Club in the Making," *Science Education* 40 (April 1956): 228–232.

53. Westinghouse Electric Corporation, "Programs and Progress Westinghouse Educational Foundation December, 1949," CBS WEF, Box "Jon Olsen's Office," Folder "WEDND—Deed w/ Trust," 1–22. The quotation appears on page 2. Westinghouse's leaders did not view most of the foundation's scholarships and fellowships as recruiting mechanisms for future employees, although the number of student recipients interviewing for jobs with their company between 1946 and 1949 had grown from 862 to 2,515. Some in the Atomic Energy Commission, meanwhile, appreciated Westinghouse's philanthropy. Edward Trapnell to Watson Davis, May 4, 1953, HUA HUG 4773.10, Box 15D, Folder "Science Service—Watson Davis."

54. Charles Johnson, Jay Moorhead, and Harriet Gormley, "Fundamental Research Vital to Industry, Talent Search Finalists Told," March 4, 1956, SIA RU 7091, Box 330, Folder 5, 1–4; Science Service, "Program of the Science Talent Institute Washington, D.C. Thursday, March 1, through Monday, March 5, 1956," SIA RU 7091, Box 330, Folder 5.

55. The foundation's support for higher education also included endowed professorships at four universities and an award in engineering education to help academic salaries compete with industry. It forged ties with the American Association for the Advancement of Science by establishing the George Westinghouse Science Writing Awards. From 1954 to 1958, moreover, the Westinghouse Educational Foundation committed an additional four million dollars to American institutions of higher education. Westinghouse Electric Corporation, "Programs and Progress Westinghouse Educational Foundation December, 1949," 1–22; Westinghouse Educational Foundation, "Exhibit I: Westinghouse Scholarship and Fellowship History," [c. 1958], CBS WEF, Box "Jon Olsen's Office," Folder "W. Educational Foundation: Historical," 1–5.

56. The company also circulated a pamphlet to schools authored by Westinghouse's chairman, A. R. Robertson, titled "Big Business Is Good Business." "Announcing 6 New Charts on Nuclear Physics," *Science News Letter* 56 (September 10, 1949): 162; "You'll Want to See and Show...'Adventures in Research'," *Science News Letter* 52 (November 29, 1947): 338; "Helping America 'Discover' Tomorrow's Leaders in Science," *Science News Letter* 51 (April 19, 1947): 242; "We Invite Your Aid in the Annual Science Talent Search," *Science News Letter* 52 (October 18, 1947): 243; "From Grammar School through College...Here's help for America's Future Scientists," *Science News Letter* 52 (September 20, 1947): 178.

57. "Points for Talk on School Program," [c. 1951], CBS WEF, Box "Ruch Docs, 2 of 2, Science Talent Search," Folder "Education—Westinghouse School Services Program," 1–6. The quotations appear on pages 1 and 5. Westinghouse commissioned a study in

1951 to determine whether the implementation of its school materials led secondary school students in one school district to view the company more favorably relative to its main competitors over the course of the academic year. School Service, Westinghouse Electric Corporation, "How Westinghouse School Service Materials affect the Opinions of Secondary School Students Grades 7 to 12," [1951], CBS WEF, Box "Ruch Docs, 2 of 2, Science Talent Search," Folder "Education—Westinghouse School Services Materials," 1–11.

58. "Points for Talk on School Program," 4–5. Westinghouse also continued to place full-page advertisements in issues of *Science News Letter*, some of which touted the company's advances in atomic research for weapons and warships. Other advertisements promoted its ongoing quest for rural electrification as well as its growing number of scholarship and internship programs. "He Advanced the Bombing of Hiroshima by at least a Year!" *Science News Letter* 50 (November 16, 1946): 306; "Research...Power...Better Living," *Science News Letter* 56 (October 18, 1949): 226; "New Lamp Gives a Bargain in Sunshine," *Science News Letter* 58 (November 11, 1950): 317; "In Recognition of a Job Well Done," *Science News Letter* 59 (April 14, 1951): 226; "Co-operation with Schools to Stimulate Interest of Graduate Students," *Science News Letter* 60 (October 6, 1951): 210.

59. Westinghouse Electric Corporation, "In Furtherance of the Public Welfare: The Westinghouse Educational Foundation, Its Purpose and Achievements," [c. 1953], CBS WEF, Box "Jon Olsen's Office," Folder "WEDND—Deed w/ Trust," 1.

60. Westinghouse Educational Foundation, "The Westinghouse Educational Foundation Announces a Five-year Program of Support to Higher Education," [c. 1954], CBS WEF, Box "Ruch Docs, 2 of 2, Science Talent Search," Folder "Education—Scholarships," 4–5.

61. Westinghouse Electric Corporation, "In Furtherance of the Public Welfare," 13; "Education Best Bet to Avert Catastrophe," *Westinghouse News* 9 (June 15, 1954): 2.

62. Westinghouse Educational Foundation, "Foundation Policy and Practice," [1959], CBS WEF, Box "Jon Olsen's Office," Folder "W. Educational Foundation: Historical," 1–4; "Science Talent Search Scholarships Tripled," *Science News Letter* 72 (August 31, 1957): 133; Westinghouse Educational Foundation, "Exhibit II: Scholarship and Fellowship Support of Other Companies," [c. 1958], CBS WEF, Box "Jon Olsen's Office," Folder "WEDF Scholarships: General," 1–3.

63. "Science Talent Institute," *Science News Letter* 49 (March 9, 1946): 150–151.

64. *Adventures in Science*, February 28, 1948, SIA RU 7091, Box 394, Folder 9, 2–6. See also "Science for Its Own Sake," *Science News Letter* 51 (March 15, 1947): 164.

65. Gwilym A. Price, "From Many Walks of Life," *Science News Letter* 53 (March 13, 1948): 166–167. The quotation appears on page 167.
66. Harlow Shapley, "To the Science Talent Winners," February 28, 1946, HUA HUG 4773.10, Box 5A, Folder "Science Talent Scholarship Award Remarks, 1946, 1948, 1950."
67. "Science Talent Institute," *Science News Letter* 53 (March 6, 1948): 149.
68. *Adventures in Science*, March 5, 1949, SIA RU 7091, Box 396, Folder 28, 9. See also "Young Scientists Asked to Develop Antibiotics," *Science News Letter* 49 (March 16, 1946): 167.
69. E. U. Condon, "Science and Our Future," *Science News Letter* 49 (March 16, 1946): 163–166.
70. "Science and Our Future," *Appendix to the Congressional Record, Proceedings and Debates of the 79th Congress, Second Session, Volume 92, Part 9* (Washington, DC: United States Government Printing Office, 1946), A1182–A1184; Wang, *American Science in an Age of Anxiety*, 21.
71. *Adventures in Science*, March 1, 1947, SIA RU 7091, Box 393, Folder 9, 5.
72. Basil O'Connor, "Science and Humanity," March 7, 1949, SIA RU 7091, Box 313, Folder 6, 4 & 6.
73. Harlow Shapley, "Draft of Remarks of Harlow Shapley to the Forty Winners of the Seventh Annual Science Talent Search, Hotel Statler, March 2, 1948," 1948, HUA HUG 4773.10, Box 5A, Folder "Science Talent Scholarship Award Remarks—1946, 1948, 1950." See also W. W. Waymack, "'Much Shall Be Required,'" *Science News Letter* 53 (March 6, 1948): 154–156.
74. J. Robert Oppenheimer, "The Encouragement of Science," March 6, 1950, CBS WEF, Box "Ruch Docs, 2 of 2, Science Talent Search," Folder "Westinghouse," 1–10. The quotations appear on pages 5 and 9, respectively.
75. M. H. Trytten, "Science Talent Problems," *Science News Letter* 49 (March 16, 1946): 170, 172, 174. The quotations appear on page 170. See also "Western Culture Minority," *Science News Letter* 51 (March 8, 1947): 150.
76. Truman quoted in Margaret E. Patterson, "Science Talent Search Light," December 1949, SIA RU 7091, Box 444, Folder 28, 2.
77. "Search for Science Talent," *Science News Letter* 58 (October 14, 1950): 247. See also, Wadsworth Likely, "Discover Scientists of the Future," *Science News Letter* 60 (October 6, 1951): 218–219; "U.S. Short 130,000 Experts," *Science News Letter* 60 (November 24, 1951): 326; Watson Davis, "Science Outlook for 1952," *Science News Letter* 61 (January 5, 1952): 10.
78. Donald A. Quarles, "Cultivating Our Science Talent—Key to Long-term Security," *The Scientific Monthly* 80 (June 1955): 352–355. The quotation appears on page 352.

79. Henry DeWolf Smyth, "Basic Science Needed," *Science News Letter* 59 (March 10, 1951): 154–156.
80. Walter G. Whitman, "Science for Defense," *Science News Letter* 61 (March 8, 1952): 154–155.
81. Caryl P. Haskins, "Address by Caryl P. Haskins," March 5, 1956, SIA RU 7091, Box 330, Folder 5, 1–8. The quotations appear on pages 4 and 5.
82. Charles Johnson, Jay Moorhead, and Harriet Gormley, "Naval Ordnance Laboratory Reveals Hypervelocity Gun," March 5, 1956, RU 7091, Box 330, Folder 5. Some talent search finalists occasionally expressed their distaste for the military applications of scientific knowledge. In 1953, Shapley communicated that several had objected to Leonard Carmichael's address about the nation's shortage of scientists, because the "ultimate value of STSers will be a great deal more than just military manpower." Others particularly disliked the scheduled trip to an air force base. As a whole, however, finalists in 1953 rated their trip to the Naval Ordnance Laboratory favorably. One in particular noted that the visit "strengthened our faith in government's help for science." See Margaret Patterson, "Evaluation of STI-53 by STS-W-53," 1953, HUA HUG 4773.10, Box 15D, Folder "Harlow Shapley—Science Service Business," 1–8; Harlow Shapley to Margaret E. Patterson, July 20, 1953, HUA HUG 4775.10, Box 15D, Folder "Science Service—1951–1955."
83. Nancy Smith Midgette, *To Foster the Spirit of Professionalism: Southern Scientists and State Academies of Science* (Tuscaloosa, AL: University of Alabama Press, 1991), 184–198; C. L. Baker, *History of Academy Conference, 1926–1970* (1971) (ERIC Document, ED 133233), 20–23; Glenn W. Blaydes, "Minutes of the A.A.A.S. Academy Conference," December 27, 1946, AAAS Archives, T-4-4, Box 2-1-1, Folder "Minutes of the AAAS Academy Conference – December 27, 1946," 1–5.
84. E. C. L. Miller, "What Can and Should a State Academy of Science Do?" in Blaydes, "Minutes of the A.A.A.S. Academy Conference," 2–4. The quotation appears on page 2.
85. Herbert Zim, "Junior Scientists Assembly," *The Science Teacher* 14 (February 1947): 24–27, 42; Margaret E. Patterson, "Junior Scientists Assemble in Chicago," *The Science Teacher* 15 (February 1948): 27–29; "Young Scientists Tell How They Got Started," *Science News Letter* 53 (January 3, 1948): 9; Keith C. Johnson, "Third Annual Junior Scientists' Assembly," *The Science Teacher* 16 (February 1949): 28–31.
86. John W. Thomson, Jr., "The First Year of the Wisconsin Junior Academy of Science, 1944–1945," in *Transactions of the Wisconsin Academy of Sciences, Arts and Letters*, ed. Banner Bill Morgan, 347–352 (Madison, WI: Atwood & Culver, 1945); John W. Thomson, Jr., "How the Junior Academies of Science Operate," *The Scientific*

Monthly 64 (April 1947): 327–336; American Association for the Advancement of Science, "Academy supplement to the Operational Guide American Association for the Advancement of Science," [c. 1950], SIA RU 7091, Box 369, Folder 1, 1–2; "Young Scientists Wanted," *Science News Letter* 54 (November 6, 1948): 302; Science Clubs of America, "Your Work," July 1947, Waldo LaSalle Schmitt Papers, 1907–1978, housed at the Smithsonian Institution Archives, in Washington, DC (hereafter referred to as SIA RU 7231), Box 51, Folder 13, 1–19; "Program of the Winter Meeting," *Indiana Academy of Science Proceedings* 55 (1945): vii–xxiii; Philip N. Powers, "The Changing Manpower Picture," *The Scientific Monthly* 70 (March 1950): 165–171.

87. Science Clubs of America, "Your Work," November 1947, SIA RU 7231, Box 51, Folder 13, 1–19; Watson Davis, "The National Program for Science Talent," December 30, 1949, SIA RU 7091, Box 446, Folder 37, 1–3.

88. Leland H. Taylor, "Report of the Secretary of the Academy Conference," 1951, AAAS Records, Box S-14-1, Folder "Academy Conference (– 1954)," 2–3.

89. "Young Scientists Wanted," 302; "Junior Academy News," *Transactions of the Kansas Academy of Science* 53 (June 1950): 127; "Report of the Junior Academy Committee," *Proceedings of the West Virginia Academy of Science* 21 (December 1949): 13; Margaret E. Patterson, "State Science Talent Searches, 1949–1950," 1949, SIA RU 7091, Box 446, Folder 37, 1–25; "High School Relations Committee," *Iowa Academy of Science Proceedings* 53 (April 1946): 27; Margaret E. Patterson, "Science Club Members Profit from State and National Cooperation," *The Science Teacher* 14 (October 1947): 124–127.

90. Science Clubs of America, "Your Work," July 1947, 4; Patterson, "Science Clubs Profit from State and National Cooperation," 125–126; "Report of Committee on Science Talent Search Scholarships," *Transactions of the Illinois State Academy of Science* 40 (May 1947): 239; Patterson, "State Science Talent Searches," 8–9.

91. John Xan, "Alabama State Science Talent Search," *The Journal of the Alabama Academy of Science* 20 (December 1948): 96.

92. Kassner subsequently visited with science teachers and principals of African–American schools in the hopes of fostering more science clubs and a parallel junior academy. He discovered that African–American schools in Alabama already had such an organization in place. Kassner nonetheless offered the state academy's help in assisting any science clubs that sought guidance. Patterson, "State Science Talent Searches," 4–5; "Minutes of the Thirteenth Annual Meeting of the Alabama Junior Academy of Science," *Alabama Academy of Science Journal* 19 (December 1947): 90; Science Clubs of America, "Your Work," July 1947, 2; Science Clubs of America,

"Your Work," November 1947, 2; Xan, "Alabama State Science Talent Search," 96; "Minutes of the Fifteenth Annual Meeting of the Alabama Junior Academy of Science, May 6, 1949," *The Journal of the Alabama Academy of Science* 21 (February 1952): 71–72; Midgette, *To Foster the Spirit of Professionalism*, 164–167, 185–188, 191–193.

93. Roughly one thousand junior and senior high school students exhibited their projects at the seventh annual science fair in the spring of 1953. In 1955, prizewinners received funds to travel to the National Science Fair in Cleveland.
94. A. T. McPherson, E. H. Walker, and M. A. Mason, "Report to Board of Managers, Washington Academy of Sciences from Special Committee on a Junior Academy of Science," April 30, 1952, Washington Academy of Sciences Records, Smithsonian Institution Archives (hereafter SIA RU 7099), Box 10, Folder "Washington Academy of Sciences," 1–3; A. T. McPherson, "Report on Encouragement of Science Talent," October 13, 1952, SIA RU 7099, Box 10, Folder "Washington Academy of Sciences, 1888–1968," 1–3; A. T. McPherson, "Committee on the Encouragement of Science Talent," 1953, SIA RU 7099, Box 10, Folder "Washington Academy of Sciences, 1888–1968," 1–2; A. T. McPherson, "Washington Academy of Sciences Committee on the Encouragement of Science Talent Report to the Board of Managers," May 17, 1955, SIA RU 7099, Box 10, Folder "Washington Academy of Sciences, 1888–1968"; "Washington Junior Academy of Sciences," 1957, SIA RU 7099, Box 10, Folder "Washington Academy of Sciences, 1888–1968"; "Activities of the Joint Board on Science Education," *Journal of the Washington Academy of Sciences* 48 (February 1958): 63–66.
95. "Virginia Academy of Science," [c. 1945], AAAS Records, P-14-1 (Box 2), Folder "Virginia Acad. Sci. (Affil. Acad. Sci.)."
96. "Report of the High School Relations Committee," *Iowa Academy of Science Proceedings* 62 (April 1955): 43; "Report of the High School Relations Committee of 1953," *Iowa Academy of Science Proceedings* 60 (April 1953): 37.
97. Tennessee's junior academy had been organized in 1941, and by 1955, it encouraged any interested secondary student to attend and even deliver papers in the senior academy program. One top boy and girl "judged by a committee to have made the best contribution to the program" received AAAS memberships among other prizes. "Talent Searches—State and National," *Journal of the Tennessee Academy of Science* 25 (October 1950): 298–307; W. W. Wyatt, "The Tennessee Junior Academy of Science," *Journal of the Tennessee Academy of Science* 30 (April 1955): 141–150.
98. W. H. Ward, "South Carolina Junior Academy of Science," *Bulletin of the South Carolina Academy of Science* 15 (1953): 4; Leland H.

Taylor, "Report of the Secretary of the Academy Conference," 2–3; "Report of the Secretary of the Academy Conference—1953," March 17, 1953, AAAS Records, Box S-14-1, Folder "Academy Conference (—1954)," 6–7, 9–11.

99. "Junior Academy of Science Conference Navy Pier Feb. 15–17, 1957," AAAS Archives, Box S-14-1, Folder "The Academy Conference Nov. 28, 1955," 3; "Report on Junior Academies of Science Conference," [1957], AAAS Records V-9-6, Box 5, Folder "Junior Academies," 1–7; Baker, *History of Academy Conference 1926–1970*, 55–56.

100. The American Institute struggled to cover costs for this event and was unable to hold a science fair again for several years. *117th Annual Report of the American Institute of the City of New York for the Year Ending December 31st, 1944* (New York, 1945), NYHS AIR Box 401, Folder 1; *119th Annual Report of the American Institute of the City of New York for the Year Ending December 31st, 1946* (New York, 1947), NYHS AIR, Box 401, Folder 1, 1–3; *120th Annual Report of the Board of Trustees to the members of the American Institute of the City of New York for the Year Ending December 31st, 1947* (New York, 1948), NYHS AIR, Box 401, Folder 1, 2; *121st Annual Report of the Board of Trustees to the Members of the American Institute of the City of New York for the Year Ending December 31st, 1948* (New York, 1949), NYHS AIR, Box 401, Folder 1.

101. *125th Annual Report of the Board of Trustees to the Members of the American Institute of the City of New York* (New York, 1953), NYHS AIR, Box 401, Folder 1, 1–2; "16th School Science Fair Saturday December 6, 1952, held by the American Institute of the City of New York," [1952], NYHS AIR, Box 401, Folder 1, 1–13; *130th Annual Report of the Board of Trustees to the members of the American Institute of the City of New York* (New York, 1958), NYHS AIR, Box 401, Folder 1, 1–6.

102. "State Science Fairs and Searches: A Report to Science News Letter Readers," *Science News Letter* 52 (July 26, 1947): 61; Norman R. D. Jones, "A Science Fair—Its Organization and Operation," *The Science Teacher* 15 (February 1949): 26–27; Hamilton Lyon, "Tenth Annual School Science Fair," *Pennsylvania School Journal* 98 (October 1949): 64; Emily K. Jones, "Science Fair Evokes Keen Interest," *School Activities* 27 (April 1956): 246–247; Carl F. Perry, "High School Science Fairs Are Worth the Effort," *School and Community* 42 (March 1956): 14–15.

103. "A Keyhole Look at Science Fairs," *The Science Teacher* 23 (November 1956): 328–329; "Science Fairs," April 30, 1957, SIA RU 7091, Box 401, Folder 101; Shirley Moore, "Science Fairs Grow Up," *Science News Letter* 73 (May 3, 1958): 282–283; Charles Frederick Beck, Jr., "The Development and Present Status of School Science Fairs" (unpublished diss., University of Pittsburgh, 1957), 35, 67;

"Today's Scientists of Tomorrow," *Science News Letter* 70 (October 13, 1956): 234-235.
104. Beck, Jr., "The Development and Present Status of School Science Fairs," 36; "Science Fairs Participating in the Seventh National Science Fair," 1956, SIA RU 7091, Box 402, Folder 38; "Hatching Eggs Wins Trip to National Science Fair," May 1, 1957, RU 7091, Box 401, Folder 101.
105. "First National Science Fair to Be Held," *Science News Letter* 57 (April 22, 1950): 245; "Skeletons, Lightning Generator Exhibited by Young Scientists," [1952], SIA RU 7091, Box 365, Folder 15, 1-2.
106. "Science Service Announces a New Local Contest for Newspaper Sponsorship Open to All Boys and Girls of High School Age Winners to Enter National Science Fair," 1950, SIA RU 7091, Box 365, Folder 15, 1-4; Watson Davis, "Science Fairs," 1951, SIA RU 7091, Box 444, Folder 20, 2-3.
107. Margaret E. Patterson, "Opportunity for Nation's Junior Scientists," *The Science Teacher* 17 (October 1950): 124-125. The quotation appears on page 124.
108. "Suggested Story Announcing Your Newspaper's Sponsorship of Science Fair," 1951, SIA RU 7091, Box 365, Folder 15, 1.
109. *Adventures in Science*, May 2, 1953, SIA RU 7091, Box 400, Folder 6, 4.
110. Both are quoted in "Fifth National Science Fair Purdue University Lafayette, Indiana, May 13-15, 1954," *Science News Letter* 64 (November 28, 1953): 337.
111. "The Ninth National Science Fair Conducted by Science Service in Cooperation with 150 Affiliated Science Fairs," May 1958, SIA RU 7091, Box 401, Folder 35, 8.
112. The Oak Ridge Institute of Nuclear Studies hosted the fourth National Science Fair in 1953, where the 70 student finalists competed for prizes and met with atomic scientists and engineers. Beginning in 1956, the American Medical Association (AMA) arranged for two students with the best exhibits on medical research, general health, or physical fitness to present their findings at its annual meeting. The AMA also encouraged its affiliated state and county medical societies to sponsor local and regional high school science fairs. Dewey E. Large, "Science Fair Information," *The Mathematics Teacher* 48 (January 1955): 57; Beck, Jr., "The Development and Present Status of School Science Fairs," 25-26; *Adventures in Science*, May 2, 1953, SIA RU 7091, Box 400, Folder 6, 1-2; "[Press Release Announcing Third Science Fair Winners]," May 9, 1952, SIA RU 7091, Box 365, Folder 15, 1-6; "Winning Science Projects," *Science News Letter* 63 (May 23, 1953): 326-327; "Seventh National Science Fair Municipal Auditorium Oklahoma City, Oklahoma May 10-12, 1956," *Science News Letter* 68 (October

22, 1955): 258; "AMA to Participate in National Science Fair," *Science News Letter* 69 (January 14, 1956): 18; "Medical Association Doubles Fair Awards," *Science News Letter* 69 (February 11, 1956): 83; John A. Yarbrough, "North Carolina Science Fairs and the North Carolina Academy of Science," *The High School Journal* 39 (February 1956): 270–274.

113. "Five Nobel Prize Winners to Greet Science Fair Finalists," May 12, 1951, SIA RU 7091, Box 398, Folder 10, 1–2; "Junior Scientists Converge in Washington for Science Fair," [1952], SIA RU 7091, Box 365, Folder 15, 1–2; "Fourth National Science Fair American Museum of Atomic Energy, Oak Ridge, Tenn.," 1953, SIA RU 7091, Box 400, Folder 5, 1–2; "Learn Russian, Get Ph.D.," *Science News Letter* 71 (May 18, 1957): 306.

114. "Los Angeles to Be Young Scientists' Capital," April 16, 1957, SIA RU 7091, Box 401, Folder 101, 1–2.

115. Margaret Patterson and Joseph Kraus, "A Visit to the National Science Fair," [c. 1956], SIA RU 7091, Box 402, Folder 57, 2.

116. Shirley Moore, "Laboratories Train Scientists," *Science News Letter* 73 (March 29, 1958): 202–203; "[National Science Fair Press Release]," April 30, 1957, SIA RU 7091, Box 401, Folder 101, 3.

117. Watson Davis, "The Role of the Science Writer—Interpretation of Scientific Research," in *Science and Society: A Symposium...to Consider the Interrelationships of Science and Man*, ed. Robert F. Irvin, [39–41] (Notre Dame, IN: University of Notre Dame, 1952).

118. "Address by Dr. Alan T. Waterman, Director, National Science Foundation," May 10, 1952, SIA RU 7091, Box 365, Folder 15, 2–3; Beck, Jr., "The Development and Present Status of School Science Fairs," 44.

119. "Science Fairs Aid Nation," *Science News Letter* 59 (April 21, 1951): 246.

120. "Third National Science Fair Smithsonian Institution Washington, D.C. May 8, 9, and 10, 1952," May 1, 1952, SIA RU 7091, Box 365, Folder 15, 2; "Science Manpower Conference Luncheon Friday May 9, 1952," [1952], SIA RU 7091, Box 399, Folder 4; *Adventures in Science*, September 12, 1953, SIA RU 7091, Box 400, Folder 66. See also President Eisenhower's congratulations of the fair participants via telegram in 1958: "Science Fair Winners," *Science News Letter* 73 (May 24, 1958): 327.

121. Glenn O. Carter, *127th Annual Report of the Board of Trustees to the Members of the American Institute of the City of New York* (New York, 1955), NYHS AIR, Folder 1, Box 401, 1.

122. "The Ninth National Science Fair Conducted by Science Service in Cooperation with 150 Affiliated Science Fairs Ballenger Field House Flint Junior College Flint, Michigan," 1958, SIA RU 7091, Box 401, Folder 35, 1.

123. Andrew Hartman, *Education and the Cold War: The Battle for the American School* (New York: Palgrave Macmillan, 2008), 198.
124. Ibid., 64.
125. Watson Davis, "High School Science Fairs in the Fields of Mathematics and Physical Sciences," June 23, 1955, SIA RU 7091, Box 402, Folder 57, 1. See also "Fourth Science Fair," *Science News Letter* 63 (March 21, 1953): 182.
126. Science Clubs of America, "How You can Search for Science Talent: A Book of Facts about the Ninth Annual Science Talent Search for Westinghouse Scholarships," [1949], SIA RU 7091, Box 444, Folder 6, 2–11; Harold A. Edgerton and Steuart Henderson Britt, "The Science Talent Search," *Occupations* 22 (December 1943): 177–180.
127. H. A. Edgerton and S. H. Britt, "The Annual Science Talent Search for the Westinghouse Scholarships," *Transactions of the New York Academy of Sciences* 11 (February 1949): 118–120; Harold A. Edgerton, Steuart Henderson Britt, and Ralph D. Norman, "Later Achievements of Male Contestants in the First Annual Science Talent Search," *American Scientist* 36 (July 1948): 403–414.
128. Banesh Hoffman, "Some Remarks Concerning the 'First Annual Science Talent Search,'" *American Scientist* 31 (July 1943): 255–265; Paul F. Brandwein, "The 'Science' Talent Search," *Science* 101 (February 2, 1945): 117; Paul F. Brandwein, "Some Comments on the Annual Science Talent Search," *Science Education* 28 (February 1944): 47–49.
129. Edgerton is quoted on page 2 in Margaret Patterson, "Science Talent Search Light," February 1948, SIA RU 7091, Box 446, Folder 20; Harold A. Edgerton and Steuart Henderson Britt, "Further Remarks Regarding the Science Talent Search," *American Scientist* 31 (July 1943): 263–265; Harold A. Edgerton and Steuart Henderson Britt, "Science Talent in American Youth," *Science* 101 (March 9, 1945): 247–248; Harold A. Edgerton, Steuart Henderson Britt, and William B. Lemmon, "Reliability of Anecdotal Material in the First Annual Science Talent Search," *Journal of Applied Psychology* 31 (August 1947): 413–424.
130. Patterson, "Science Talent Search Light," December 1949, 60–61.
131. Davis is quoted on page iv in Harold A. Edgerton, *Science Talent: Its Early Identification and Continuing Development* (New York: Richardson, Bellows, Henry & Company, 1961).
132. Watson Davis contended in 1942 that "the question of race, color, or creed was not in any way specified, so that we only by accident know of Negro participants." Edgerton and Britt similarly emphasized that judges had no knowledge of an entrant's race or religion until after the 300 winners and honorable mentions had been selected. Watson Davis to G. Lake Imes, July 7, 1942, SIA RU 7091, Box 236, Folder 7; Harold A. Edgerton, Steuart Henderson Britt, and Ralph D. Norman, "Physical Differences between Ranking

and Non-ranking Contestants in the First Annual Science Talent Search," *American Journal of Physical Anthropology* 5 (December 1947): 435–452.
133. Edgerton, Britt, and Norman, "Physical Differences between Ranking and Non-ranking Contestants in the First Annual Science Talent Search," 441.
134. Harold A. Edgerton and Steuart Henderson Britt, "The Science Talent Search in Relation to Educational and Economic Indices," *School and Society* 62 (March 9, 1946): 172–175.
135. This claim derives from my analysis of demographic data and published photographs of all 680 of the Science Talent Search winners from 1942 to 1958.
136. Girls also scored lower than boys on the science aptitude examination from 1942 to 1944. Harold A. Edgerton and Steuart Henderson Britt, "Sex Differences in the Science Talent Test," *Science* 100 (September 1, 1944): 192–193. The quotation appears on page 193.
137. Available data for honorable mentions include 1942–1945, 1952, and 1956–1958. Data are compiled from the following sources: Science Service, *Youth Looks at Science and War: A Collection of Essays by the Washington Trip Winners of the First Annual Science Talent Search Conducted by Science Clubs of America* (Washington, DC: Penguin Books, 1942), 104–108; Science Clubs of America, *Science and the Future: Essays of the Winners of the Westinghouse Science Scholarships in the Second Annual Science Talent Search* (Washington, DC: Science Clubs of America, 1943), 107–111; Science Clubs of America, *Scientists of Tomorrow: Essays of the Winners of the Westinghouse Science Scholarships in the Third Annual Science Talent Search* (Washington, DC: Science Service, 1944), 120–124; Science Clubs of America, *Scientists of Tomorrow: Essays of the Winners of the Westinghouse Science Scholarships in the Fourth Annual Science Talent Search* (Washington, DC: Science Service, 1945), 118–123; Science Clubs of America, "The Winners and Honorable Mentions in the Eleventh Annual Science Talent Search," [1952], SIA RU 7091, Box 401, Folder 4, 5–10; Science Clubs of America, "The Winners and Honorable Mentions in the 15th Annual Science Talent Search," [1956], SIA RU 7091, Box 402, Folder 56, 5–10; Science Clubs of America, "The Winners and Honorable Mentions in the 16th Annual Science Talent Search," [1957], SIA RU 7091, Box 402, Folder 6, 5–10; Science Clubs of America, "The Winners and Honorable Mentions in the 17th Annual Science Talent Search," [1958], SIA RU 7091, Box 401, Folder 39, 5–14.
138. Marina Prajmovsky met her husband at Yale Medical School and the couple interned in 1951 at the same hospital in Michigan. Margaret Grace, meanwhile, anticipated transitioning from a housewife to teaching psychology at Barnard College. Margaret E. Patterson,

"Science Talent Search Light," August 1951, SIA RU 7091, Box 444, Folder 28, 17; Patterson, "Science Talent Search Light," February 1948, 14, 49.
139. Patterson, "Science Talent Search Light," February 1948, 21.
140. Patterson, "Science Talent Search Light," December 1949, 12.
141. Ibid.
142. Edgerton, *Science Talent*, 49–50.
143. Linda Eisenmann, *Higher Education for Women in the Postwar Era, 1945–1965* (Baltimore: Johns Hopkins University Press, 2006), 11–42; Patterson, *Grand Expectations*, 363–369; Margaret W. Rossiter, *Women Scientists in America: Before Affirmative Action 1940–1972* (Baltimore: Johns Hopkins University Press, 1995); Sevan G. Terzian, "*Science World*, High School Girls, and the Prospect of Scientific Careers, 1957–1963," *History of Education Quarterly* 46 (Spring 2006): 73–99.
144. Edgerton, *Science Talent*, 10–11.
145. Patterson, "Science Talent Search Light," February 1948, 10.
146. Ibid., 18.
147. Rossiter, *Women Scientists in America*, 123–128, 137–141. See also Eisenmann, *Higher Education for Women in Postwar America*, 13–19.
148. Lyle described her exclusively domestic role as educational: "There are as many things to learn about cooking, cleaning, and sewing as any chemistry course." Patterson, "Science Talent Search Light," December 1949, 17.
149. Edgerton acknowledged that the 20 married women he interviewed tended to prize domestic responsibilities and that "child bearing limited their professional work," which disrupted their career paths. Yet he concluded that "there is a general buoyant attitude that balance in the dual role is possible." Edgerton, *Science Talent*, 9–10, 11, 53; Patterson, "Science Talent Search Light," August 1951, 13.
150. Kenneth E. Brown and Ellsworth S. Obourn, *Offerings and Enrollments in Science and Mathematics in Public High Schools 1958, Bulletin 1961, No. 5* (Washington, DC: U.S. Government Printing Office, 1961), 36–38, 43.
151. Steven Mintz, *Huck's Raft: A History of American Childhood* (Cambridge, MA: Harvard University Press, 2004), 284.
152. Data are compiled from the following sources: "Science Talent Search Washington Trip Winners," *Science News Letter* 41 (June 27, 1942): 405; Science Clubs of America, *Science and the Future*, 105–106; Science Clubs of America, *Scientists of Tomorrow* [1944], 118–119; Science Clubs of America, *Scientists of Tomorrow* [1945], 117–118; "40 Winners Listed," *Science News Letter* 49 (February 2, 1946): 70; "40 Winners to Compete," *Science News Letter* (January 25, 1947): 54; "Seventh Annual Science Talent Search Washington

Trip Winners" [1948], SIA RU 7091, Box 394, Folder 7, 1–4; "How You Can Search for Science Talent," [1949], 10–11; "STS Winners Are Selected," *Science News Letter* 57 (February 11, 1950): 86; "STS Winners Selected," *Science News Letter* 59 (February 10, 1951): 86; Science Clubs of America, "The Winners and Honorable Mentions in the Eleventh Annual Science Talent Search," 3–4; "Supplementary Information Twelfth Annual Science Talent Search," [1953], SIA RU 7091, Box 402, Folder 5, 1–21; "12th Annual Science Talent Search 1953," [1953], SIA RU 7091, Box 401, Folder 5; Science Clubs of America, "How you can Search for Science Talent: A Book of Facts about the Fourteenth Science Talent Search for Westinghouse Science Scholarships," [1954], SIA RU 7281, William F. Foshag Collection, 1923–1965, and Undated, Box 4, Folder "Science Clubs of America 1955," 10–11; "The Fourteenth Annual Science Talent Search," [1955], SIA RU 7091, Box 330, Folder 5; "15th Science Talent Search," [1956], RU 7091, Box 402, Folder 49; Science Clubs of America, "The Winners and Honorable Mentions in the 16th Annual Science Talent Search," 3–4; Science Clubs of America, "The Winners and Honorable Mentions in the 17th Annual Science Talent Search," [1958], SIA RU 7091, Box 401, Folder 39, 3–4.
153. Patterson, *Grand Expectations*, 325; Mintz, *Huck's Raft*, 276.
154. Only three of the 12 states in the Southeast—Texas, Florida, and Louisiana—had a majority urban population according to the 1950 census. The percentage of rural residents in the remaining nine states ranged from 53.0 percent (Virginia) to 72.1 percent (Mississippi). U.S. Office of Education, *Statistics of State School Systems, 1949–1950* (Washington, DC: United States Government Printing Office, 1952), 47–49; U.S. Census Bureau, *Census of Population, 1950 Census, Volume 2, Characteristics of the Population* (Washington, DC: United States Government Printing Office, 1953).
155. Brown and Obourn, *Offerings and Enrollments in Science and Mathematics in Public High Schools 1958, Bulletin 1961, No. 5*, 33; U.S. Office of Education, "Offerings and Enrollments in High-School Subjects," 53–54.
156. Samuel W. Bloom, "The Search for Science Talent," *Science Education* 38 (April 1954): 232–236. Harold Edgerton concluded in 1961 that a high school's size constituted a salient factor: "Students who stayed in science came, on the average, from larger high schools than those who went into non-science fields. This suggests that there is a greater training potential or science sophistication in the larger schools." Edgerton, *Science Talent*, iii.
157. All but one of the top 20 high schools in the talent search from 1942 to 1990 was located in metropolitan areas. The lone exception

was an elite private boarding school, the Phillips Exeter Academy in Exeter, New Hampshire. See Phares, *Seeking—and Finding—Science Talent*, 86–88. Senior class data for the 1953 talent search winners are from "Supplementary Information Twelfth Annual Science Talent Search," [1953], 1–21.
158. James B. Conant, *The American High School Today: A First Report to Interested Citizens* (New York: McGraw-Hill Book Co., 1959), 37, 59, 73, 80 (the quotation appears on page 37); Brown and Obourn, *Offerings and Enrollments in Science and Mathematics in Public High Schools 1958*, 8–9.

Conclusion

1. Morris Meister, *Children's Science Fair of The American Institute: A Project in Science Education* (New York: The American Institute and The American Museum of Natural History, 1932), 7.
2. "Science Fair Winners," *Science News Letter* 71 (May 25, 1957): 326.
3. Alan T. Waterman, "Role of the Federal Government in Science Education," *The Scientific Monthly* 82 (June 1956): 286–293. The quotation appears on page 289.
4. Dwight D. Eisenhower, "Science in National Security: Present Strength and Future Plans," *Vital Speeches of the Day* XXIV (November 15, 1957): 66–69. The quotation appears on page 68.
5. Charles C. Cole, Jr., *Encouraging Scientific Talent: A Study of America's Able Students Who Are Lost to College and Ways of Attracting Them to College and Science Careers* (New York: College Entrance Examination Board, 1956): 184. See also Robert Douglas MacCurdy, "Characteristics and Backgrounds of Superior Science Students," *The School Review* 64 (February 1956): 67–71, and "Biographical Sketch Dr. Howard L. Bevis, Chairman National Committee for the Development of Scientists and Engineers," 1957, SIA RU 7091, Box 402, Folder 7.
6. *Adventures in Science*, February 22, 1958, SIA RU 7091, Box 401, Folder 39.
7. John L. Rury and Shirley A. Hill, *The African American Struggle for Secondary Schooling, 1940–1980: Closing the Graduation Gap* (New York: Teachers College Press, 2012).
8. John Carson, *The Measure of Merit: Talents, Intelligence, and Inequality in the French and American Republics, 1750–1940* (Princeton: Princeton University Press, 2007), 5. See also Amy Slaton, *Race, Rigor, and Selectivity in U.S. Engineering: The History of an Occupational Color Line* (Cambridge, MA: Harvard University Press, 2010).
9. Thomas J. Sugrue, "Reassessing the History of Postwar America," *Prospects* 20 (1995): 493–509.

10. Wayne J. Urban, *More Than Science and Sputnik: The National Defense Education Act of 1958* (Tuscaloosa: University of Alabama Press, 2010).
11. Hosting the inaugural White House Science Fair in 2010, for example, President Barack Obama sought to recognize publicly the achievements of extraordinary students in science with the same esteem given to professional athletes. "That's what's going to help ensure that we succeed in the next century, that we're leading the world in developing the technologies, businesses and industries of the future," the president proclaimed. He did not cite these programs for creating rational, active, and empathetic citizens in a participatory democracy. Barack Obama, "Remarks by the President at White House Science Fair," October 18, 2010. http://www.whitehouse.gov/the-press-office/2010/10/18/remarks-president-white-house-science-fair (accessed June 13, 2011); Phil Larson, "President Obama Meets with Fresh Science Talent," March 15, 2011. http://whitehouse.gov/blog/2011/03/15/president-obama-meets-fresh-science-talent (accessed June 13, 2011). See also an account of President George H. W. Bush's remarks to the Science Talent Search finalists in 1989 in Julie Ann Miller, "Science Education for Whom?" *BioScience* 39 (June 1989): 354.
12. In praising the 40 finalists of the Science Talent Search in 2009, Intel chairman, Craig Barrett, reflected the same rationale that Westinghouse had articulated in years past. "At a time when our country requires innovation to spur economic growth," Barrett declared, "it is inspiring to see such talented young people using critical thinking skills to find solutions to scientific challenges." Rachel Ehrenberg, "Science's Next Superstars Win Fortune and Fame," *Science News* 175 (March 28, 2009): 12; J. A. Miller, "Science Talent Search Future," *Science News* 153 (January 10, 1998): 20; "Science Talent Search Has New Sponsor," *Science News* 153 (March 28, 1998): 196.
13. American Association for the Advancement of Science, *Science for All Americans* (New York: Oxford University Press, 1990). The quotations are from pages xviii and xiii.
14. National Science Board, *America's Pressing Challenge—Building a Stronger Foundation. A Companion to Science and Engineering Indicators—2006* (Arlington, VA: National Science Foundation, January 2006), [iii], 2–3, 5. More obliquely and to a lesser extent, the NSB also identified a need to close achievement gaps among American youth that hindered equal educational opportunities and discouraged the active participation of citizens on issues concerning science and technology.
15. National Academy of Sciences, National Academy of Engineering, and Institute of Medicine, *Rising Above the Gathering Storm:*

Energizing and Employing America for a Brighter Economic Future (Washington, DC: National Academies Press, 2007), 3, ix, 13.
16. National Academy of Sciences, National Academy of Engineering, and Institute of Medicine, *Rising Above the Gathering Storm Revisited: Rapidly Approaching Category 5* (Washington, DC: National Academies Press, 2010), 2, 4.
17. Domestic Policy Council, *American Competitiveness Initiative: Leading the World in Innovation* (Washington, DC: Office of Science and Technology Policy, 2006), 4, 15.
18. Ibid., 13–18.
19. Andrew Jewett, "Science & the Promise of Democracy in America," *Daedalus* 132 (Fall 2003): 64–70.
20. Kenneth Prewitt, "Scientific Illiteracy and Democratic Theory," *Daedalus* 112 (Spring 1983): 49–64.

Selected Bibliography

Manuscript and Archival Materials:

American Association for the Advancement of Science, Washington, DC.
Records of the American Association for the Advancement of Science.

CBS Corporation Inc. Offices, Pittsburgh.
Westinghouse Educational Foundation Records.

Harvard University Archives, Boston.
Papers of Harlow Shapley, 1906–1966, HUG 4773.

New-York Historical Society, New York City.
American Institute of the City of New York for the Encouragement of Science and Invention Records, 1808–1983, MS 17.

Senator John Heinz History Center, Pittsburgh.
Records from the George Westinghouse Museum, 2010.0153.
Westinghouse Electric Corporation Records, 1865–2000, MSS 424

Smithsonian Institution Archives, Washington D.C.
Science Service Records, Record Unit 7091.
Waldo LaSalle Schmitt Papers, 1907–1978, Record Unit 7231.
Washington Academy of Sciences Records, Record Unit 7099.

Newspapers and Periodicals:

Amateur Scientist
New York Times
News Notes
Science News
Science News Letter
Science Observer
The American Institute Monthly
The Centurian
The March of Science
The Science Leaflet
The Westinghouse News
Westinghouse Magazine

Government Publications

Astell, Louis A., and Charles W. Odell. *High School Science Clubs. Bureau of Educational Research Bulletin No. 60.* Urbana: University of Illinois, 1932.

Brown, Kenneth E., and Ellsworth S. Obourn. *Offerings and Enrollments in Science and Mathematics in Public High Schools 1958, Bulletin 1961, No. 5.* Washington, DC: U.S. Government Printing Office, 1961.

Farrell, George E. *Miscellaneous Circular No. 85: Boys' and Girls' 4-H Club Work under the Smith-Lever Act, 1914–1924.* Washington, DC: United States Department of Agriculture, 1926.

Jessen, Carl A., and Lester B. Herlihy. *Offerings and Registrations in High-school Subjects, Bulletin 1938, No. 6.* Washington, DC: United States Department of the Interior, 1938.

"Science and Our Future," *Appendix to the Congressional Record, Proceedings and Debates of the 79th Congress, Second Session, Volume 92, Part 9.* Washington, DC: United States Government Printing Office, 1946.

U.S. Census Bureau. *Census of Population, 1950 Census, Volume 2, Characteristics of the Population.* Washington, DC: United States Government Printing Office, 1953.

U.S. Office of Education. *Statistics of State School Systems, 1949–1950.* Washington, DC: United States Government Printing Office, 1952.

U.S. Senate Subcommittee on War Mobilization. *Hearings on Science Legislation: S. 1297 and Related Bills.* Washington, DC: United States Government Printing Office, 1946.

Books, Articles, Book Chapters, Dissertations, and Theses

"A Keyhole Look at Science Fairs." *The Science Teacher* 23 (November 1956): 328–329.

"Activities of the Joint Board on Science Education." *Journal of the Washington Academy of Sciences* 48 (February 1958): 63–66.

Akard, Patrick. "Social and Political Elites." In *Encyclopedia of Sociology Volume 4*, ed. Edgar F. Borgatta and Rhonda J. V. Montgomery, 2622–2630. New York: Macmillan Reference USA, 2001.

American Association for the Advancement of Science. *Science for All Americans.* New York: Oxford University Press, 1990.

"Annual Children's Science Fair of the American Institute." *School and Society* 38 (October 14, 1933): 497.

Astell, Louis A. "Fostering Science Clubs in the High School." *Journal of Chemical Education* 6 (March 1929): 496–501.

———. "How State Academies of Science May Encourage Scientific Endeavor among High-school Students." *Science* 71 (May 2, 1930): 447.

———. "The Inspiration Which the Junior Academy of Science Has Brought to the High School Science Clubs in the State of Illinois." *School Science and Mathematics* 32 (October 1932): 748–757.

Astell, Louis A., and S. Aleta McEvoy, "Report of the Committee on High-school Science Clubs," *Transactions of the Illinois State Academy of Science* 23, no. 1 (1930): 25–30.

Baker, C. L. *A History of Academy Conference, 1926–1970.* (Memphis, TN: ERIC Document ED 133233, 1971).

Beck Jr., Charles Frederick. "The Development and Present Status of School Science Fairs." PhD diss., University of Pittsburgh, 1957.

Benowitz, Gilbert. "Science Club in the Making." *Science Education* 40 (April 1956): 228–232.

Berger, Joseph. *The Young Scientists: America's Future and the Winning of the Westinghouse.* Reading, MA: Addison-Wesley, 1994.

Bilsing, S. W. "Science Clubs in Relation to State Academies of Science." *Science Education* 18 (October 1934): 162–167.

Binder, Frederick M., and David M. Reiners. *All the Nations under Heaven: An Ethnic and Racial History of New York City.* New York: Columbia University Press, 1995.

Bloom, Samuel W. "The Search for Science Talent." *Science Education* 38 (April 1954): 232–236.

Boyd, Paul B. "The Future of the State Academy of Science." *Science* 51 (June 11, 1920): 575–580.

Boyer, Paul. *By the Bomb's Early Light: American Thought and Culture at the Dawn of the Atomic Age.* Chapel Hill: University of North Carolina Press, 1994.

Branch, Hazel Elisabeth. "The Aims and Opportunities of the Junior Academy in Kansas." *Transactions of the Kansas Academy of Science* 34 (April 1931): 27–32.

Branch, Sarah W. "Science Club; Raleigh, N. Car." *School Science and Mathematics* 31 (February 1931): 231–232.

Brandwein, Paul F. "Some Comments on the Annual Science Talent Search." *Science Education* 28 (February 1944): 47–49.

———. "The 'Science' Talent Search." *Science* 101 (February 2, 1945): 117.

Brinkley, Alan. "World War II and American Liberalism." In *The War in American Culture: Society and Consciousness during World War II*, ed. Lewis A. Erenberg and Susan E. Hirsch, 313–330. Chicago: University of Chicago Press, 1996.

Brown, JoAnne. "'A Is for Atom, B Is for Bomb': Civil Defense in American Public Education, 1948–1963." *Journal of American History* 75 (June 1988): 68–90.

Burnham, John C. *How Superstition Won and Science Lost: Popularizing Science and Health in the United States.* New Brunswick, NJ: Rutgers University Press, 1987.

Bush, Ethel. "Organizing the Biology Class into a Nature Study Club." *Science Education* 15 (November 1930): 48–53.

Caldwell, Otis W. "From the Viewpoint of the Interrelationship of National, State and Local Science Organizations and Clubs." *Science Education* 22 (February 1938): 70–71.

———."Science Essays by High School Pupils." *Science* 75 (April 8, 1932): 385–388.

———, and Edwin E. Slosson. *Science Remaking the World.* Garden City: Doubleday, Page & Company, 1924.

Carothers, George N. "A Science Fair." *Texas Outlook* 15 (April 1931): 16–20.

Carson, John. *The Measure of Merit: Talents, Intelligence, and Inequality in the French and American Republics, 1750–1940.* Princeton: Princeton University Press, 2007.

Cavanaugh, Matt. *Against Equality of Opportunity.* Oxford: Clarendon Press, 2002.

Chudacoff, Howard. *The Evolution of American Urban Society.* 2nd ed. Englewood Cliffs, NJ: Prentice-Hall, 1981.

Clowse, Barbara Barksdale. *Brainpower for the Cold War: The Sputnik Crisis and National Defense Education Act of 1958.* Westport, CT: Greenwood Press, 1981.

Clute, Willard N. "The High School Botanical Club." *School Science and Mathematics* 12 (February 1912): 147–149.

Cohen, Carl. "Democracy [Addendum]." In *Encyclopedia of Philosophy, Volume 2,* ed. Donald Borchert, 703–706. Detroit: Macmillan Reference USA, 2006.

Cohen, Ronald D. "Schooling Uncle Sam's Children: Education in the USA, 1941–1945." In *Education and the Second World War: Studies in Schooling and Social Change,* ed. Roy Lowe, 46–58. London and Washington, DC: Falmer Press, 1992.

Coit, Marjorie C. *Projects in Science and Nature Study.* New York: Department of Education of the Museum, 1931.

Cole Jr., Charles C. *Encouraging Scientific Talent: A Study of America's Able Students Who Are Lost to College and Ways of Attracting Them to College and Science Careers.* New York: College Entrance Examination Board, 1956.

Conant, James Bryant. "Science and Society in the Post-war World," *Vital Speeches of the Day* 9 (April 15, 1943): 394–397.

———. *The American High School Today: A First Report to Interested Citizens.* New York: McGraw-Hill Book Co., 1959.

Conn, Steven. *Do Museums Still Need Objects?* Philadelphia: University of Pennsylvania Press, 2010.

———. *Museums and American Intellectual Life, 1876–1926.* Chicago: University of Chicago Press, 1998.

Cowen, Ron. "'Go for It, Kid.' Looking Back on Five Decades of the Science Talent Search." *Science News* 139 (February 23, 1991): 120–123.

Creighton, Andrew L. "Democracy." In *Encyclopedia of Sociology, Volume 1,* ed. Edgar F. Borgatta and Rhonda J. V. Montgomery, 601–609. New York: Macmillan Reference USA, 2001.

Cremin, Lawrence. *The Transformation of the School: Progressivism in American Education, 1876–1957.* New York: Vintage Books, 1961.

Crook, A. R. "Council Meeting of the Illinois State Academy of Science." *Science* 52 (December 3, 1920): 533.
Cummings, Carlos Emmons. *East Is East and West Is West: Some Observations on the World's Fairs of 1939 by One Whose Main Interest Is in Museums.* Buffalo: Buffalo Museum of Science, 1940.
Cusker, Joseph Philip. "The World of Tomorrow: The 1939 New York World's Fair." PhD diss., Rutgers University, 1990.
Daniels, Norman. "Merit and Meritocracy." *Philosophy and Public Affairs* 7 (Spring 1978): 206–223.
Davis, Watson. "Federal Support for Postwar Scientific Research and Education." *The Educational Record* 26 (October 1945): 281–287.
———. "Science and the Press." *The Annals of the American Academy of Political and Social Science* 219 (January 1942): 100–106.
———. "Science Teaching and Science Clubs Now and Postwar." *School Science and Mathematics* 14 (March 1945): 257–264.
———. "The Role of the Science Writer—Interpretation of Scientific Research." In *Science and Society: A Symposium…to Consider the Interrelationships of Science and Man*, ed. Robert F. Irvin, [39–41]. Notre Dame, IN: University of Notre Dame, 1952.
DeBoer, George E. *A History of Ideas in Science Education: Implications for Practice.* New York: Teachers College Press, 1991.
Dewey, John. *Democracy and Education.* New York: Free Press, 1966.
———. *Experience and Education.* New York: MacMillan, 1963.
———. "My Pedagogic Creed." *School Journal* 54 (January 1897): 77–80.
Domestic Policy Council. *American Competitiveness Initiative: Leading the World in Innovation.* Washington, DC: Office of Science and Technology Policy, 2006.
Donahue, David. "Serving Students, Science, or Society? The Secondary School Physics Curriculum in the United States, 1930–65." *History of Education Quarterly* 33 (Fall 1993): 321–352.
Dorn, Charles. *American Education, Democracy, and the Second World War.* New York: Palgrave Macmillan, 2007.
Dunn, Fannie W. "The Place of the 4-H Club Work in the American System of Public Education." *Proceedings of the Sixty-sixth Annual Meeting of the National Education Association* 66 (1928): 508–517.
Duthie, Mary Eva. "4-H Club Work in the Life of Rural Youth." PhD diss., University of Wisconsin, 1936.
Edgerton, Harold A. *Science Talent: Its Early Identification and Continuing Development.* New York: Richardson, Bellows, Henry & Company, 1961.
Edgerton, Harold A., and Steuart Henderson Britt. "The Annual Science Talent Search for the Westinghouse Scholarships." *Transactions of the New York Academy of Sciences* 11 (February 1949): 118–120.
———. "Further Remarks Regarding the Science Talent Search." *American Scientist* 31 (July 1943): 263–265.

———. "Science Talent in American Youth." *Science* 101 (March 9, 1945): 247–248.

———. "Sex Differences in the Science Talent Test." *Science* 100 (September 1, 1944): 192–193.

———. "The First Annual Science Talent Search." *American Scientist* 31 (January 1943): 55–68.

———. "The Science Talent Search." *Occupations* 22 (December 1943): 177–180.

———. "The Science Talent Search in Relation to Educational and Economic Indices." *School and Society* 62 (March 9, 1946): 172–175.

Edgerton, Harold A., Steuart Henderson Britt, and Ralph D. Norman. "Later Achievements of Male Contestants in the First Annual Science Talent Search." *American Scientist* 36 (July 1948): 403–414.

———. "Physical Differences between Ranking and Non-ranking Contestants in the First Annual Science Talent Search." *American Journal of Physical Anthropology* 5 (December 1947): 435–452.

Edgerton, Harold A., Steuart Henderson Britt, and William B. Lemmon. "Reliability of Anecdotal Material in the First Annual Science Talent Search." *Journal of Applied Psychology* 31 (August 1947): 413–424.

Eisenhower, Dwight D. "Science in National Security: Present Strength and Future Plans." *Vital Speeches of the Day* 24 (November 15, 1957): 66–69.

Eisenmann, Linda. *Higher Education for Women in the Postwar Era, 1945–1965.* Baltimore: Johns Hopkins University Press, 2006.

Enders, Howard E. "Introductory Remarks Concerning the Origin of the Junior Academy of Science and Its Relation to the Indiana Academy of Science." *Teachers College Journal* 5 (September 1933): 146–148.

Exelby, Clyde L., and Lida Belle Gambill. *Science Club Manual.* Lansing, MI: National Club Manual Co., 1931.

Fass, Paula. *Outside In: Minorities and the Transformation of American Education.* New York: Oxford University Press, 1989.

———. *The Damned and the Beautiful: American Youth in the 1920s.* New York: Oxford University Press, 1977.

Federer Jr., Charles A. "Nation-wide Junior Science Clubs." *Science* 88 (December 2, 1938): 526.

Force, Edith R. "Special Activities of Science Students." *Education* 56 (March 1936): 438–440.

Frazier, Francis P. "General Science Club Notes." *School Science and Mathematics* 31 (March 1931): 341–344.

Giordano, Gerard. *Wartime Schools: How World War II Changed American Education.* New York: Peter Lang, 2004.

Graham, Hugh Davis. *The Uncertain Triumph: Federal Education Policy in the Kennedy and Johnson Years.* Chapel Hill: University of North Carolina Press, 1984.

Hart, David M. *Forged Consensus: Science, Technology, and Economic Policy in the United States, 1921–1953.* Princeton: Princeton University Press, 1998.

Hartman, Andrew. *Education and the Cold War: The Battle for the American School.* New York: Palgrave Macmillan, 2008.
Henry, Nelson B. *The Forty-sixth Yearbook of the National Society for the Study of Education: Part I Science Education in American Schools.* Chicago: University of Chicago Press, 1947.
Herbst, Jurgen. *The Once and Future School: Three Hundred and Fifty Years of American Secondary Education.* New York: Routledge, 1996.
Hersh, Bessie. "The School Nature League: Its History, Organization and Educational Contributions." MA thesis, Cornell University, 1940.
"High School Relations Committee." *Iowa Academy of Science Proceedings* 53 (April 1946): 27.
Hoffer, Thomas B. "Meritocracy." In *Education and Sociology Encyclopedia*, ed. David L. Levinson, Peter W. Cookson, and Alan R. Sadovnik, 435–442. New York: RoutledgeFalmer, 2002.
Hoffman, Banesh. "Some Remarks Concerning the 'First Annual Science Talent Search.'" *American Scientist* 31 (July 1943): 255–265.
Hunter, George W. "What a Science Club Can Do for a School." *School Science and Mathematics* 23 (December 1923): 817–820.
Hyman, Stanley Edgar, and St. Clair McKelway. "Onward & Upward with Business & Science: The Time Capsule." *New Yorker* 29 (December 5, 1953): 196–203.
Jackson, Philip W. "The Reform of Science Education: A Cautionary Tale." *Daedalus* 112 (Spring 1983): 143–166.
Jewett, Andrew. "Science & the Promise of Democracy in America." *Daedalus* 132 (Fall 2003): 64–70.
Johnson, George E. "Report of the Secretary." *Transactions of the Kansas Academy of Science* 35 (April 1932): 23–24.
Johnson, Keith C. "Third Annual Junior Scientists' Assembly." *The Science Teacher* 16 (February 1949): 28–31.
Jones, Emily K. "Science Fair Evokes Keen Interest." *School Activities* 27 (April 1956): 246–247.
Jones, Norman R. D. "A Science Fair—Its Organization and Operation." *The Science Teacher* 15 (February 1949): 26–27.
"Junior Academy News." *Transactions of the Kansas Academy of Science* 53 (June 1950): 127.
"Junior Science Clubs at the World's Fair." *School and Society* 48 (July 2, 1938): 10.
Kaestle, Carl. *Pillars of the Republic: Common Schools and American Society, 1780–1860.* New York: Hill & Wang, 1983.
Kaiser, David. "Cold War Requisitions, Scientific Manpower, and the Production of American Physicists during World War II." *Historical Studies in the Physical and Biological Sciences* 33, no. 1 (2002): 131–159.
Karabel, Jerome. *The Chosen: The Hidden History of Admission and Exclusion at Harvard, Yale, and Princeton.* New York: Houghton Mifflin, 2005.

Katsillis, John, and J. Michael Armer. "Education and Mobility." In *Encyclopedia of Sociology Volume 2*, ed. Edgar F. Borgatta and Rhonda J. V. Montgomery, 755–760. New York: Macmillan Reference USA, 2001.

Kennedy, David M. *Freedom from Fear: The American People in Depression and War, 1929–1945*. New York: Oxford University Press, 2005.

Kevles, Daniel J. "The National Science Foundation and the Debate over Postwar Research Policy, 1942–1945: A Political Interpretation of *Science—The Endless Frontier.*" *Isis* 68 (March 1977): 5–26.

Kinney, Harrison. "The Year of the Gifted Children." *Think* 45 (September/October 1979): 12–17.

Klopfer, Leopold E., and Audrey B. Champagne. "Ghosts of Crisis Past." *Science Education* 74 (April 1990): 133–154.

Kohlstedt, Sally Gregory. "'A Better Crop of Boys and Girls': The School Gardening Movement, 1890–1920." *History of Education Quarterly* 48 (February 2008): 58–93.

———. *Teaching Children Science: Hands-on Nature Study in North America, 1890–1930*. Chicago: University of Chicago Press, 2010.

———. "'Thoughts in Things': Modernity, History, and North American Museums." *Isis* 96 (December 2005): 586–601.

Kohlstedt, Sally Gregory, Michael M. Sokal, and Bruce V. Lewenstein. *The Establishment of Science in America: 150 Years of the American Association for the Advancement of Science*. New Brunswick: Rutgers University Press, 1999.

Krug, Edward A. *The Shaping of the American High School*. New York: Harper & Row, 1964.

———. *The Shaping of the American High School Volume 2, 1920–1941*. Madison: University of Wisconsin Press, 1972.

Kurzhals, Jane Ellen Gindelberger. "A History of the Illinois Junior Academy of Science." MA thesis, MacMurry College, 1956.

Kuznick, Peter J. *Beyond the Laboratory: Scientists as Political Activists in 1930s America*. Chicago: University of Chicago Press, 1987.

———. "Losing the World of Tomorrow: The Battle over the Presentation of Science at the 1939 New York World's Fair." *American Quarterly* 46 (September 1994): 341–373.

LaFollette, Marcel C. *Making Science Our Own: Public Images of Science, 1910–1955*. Chicago: University of Chicago Press, 1990.

———. *Science on the Air: Popularizers and Personalities on Radio and Early Television*. Chicago: University of Chicago Press, 2008.

———. "Taking Science to the Marketplace: Examples of Science Service's Presentation of Chemistry during the 1930s." *HYLE—International Journal for Philosophy of Chemistry* 12 (June 2006): 67–97.

Large, Dewey E. "Science Fair Information." *The Mathematics Teacher* 48 (January 1955): 57.

Larson, Phil. "President Obama Meets with Fresh Science Talent." March 15, 2011. http://whitehouse.gov/blog/2011/03/15/president-obama-meets-fresh-science-talent (accessed June 13, 2011).

Leff, Mark H. "The Politics of Sacrifice on the American Home Front in World War Two." *Journal of American History* 77 (March 1991): 1296–1318.
Lewenstein, Bruce V. "'Public Understanding of Science' in America, 1945–1965." PhD diss., University of Pennsylvania, 1987.
———. "The Meaning of 'Public Understanding of Science' in the United States after World War II." *Public Understanding of Science* 1 (January 1992): 45–68.
Lewis, J. Arthur. "Experiences with Science Clubs." *School Science and Mathematics* 23 (October 1923): 624–629.
Littell, Robert. "Staying after School for Fun." *Reader's Digest* 31 (October 1937): 14–16.
Livingston, Burton E. "The Atlantic City Meeting of the American Association for the Advancement of Science and Associated Studies." *Science* 77 (February 3, 1933): 129.
Lussenhop, Raymond. "The Organization of a Science Club." *School Science and Mathematics* 24 (October 1924): 727–730.
Lyon, Hamilton. "Tenth Annual School Science Fair." *Pennsylvania School Journal* 98 (October 1949): 64.
MacCurdy, Robert Douglas. "Characteristics and Backgrounds of Superior Science Students." *The School Review* 64 (February 1956): 67–71.
Marchand, Roland, and Michael L. Smith. "Corporate Science on Display." In *Scientific Authority and Twentieth Century America*, ed. Ronald Walters, 148–184. Baltimore: Johns Hopkins University Press, 1997.
McGerr, Michael. *A Fierce Discontent: The Rise and Fall of the Progressive Movement in America, 1870–1920*. New York: Free Press, 2003.
McNamee, Stephen J., and Robert K. Miller, Jr. *The Meritocracy Myth*. Lanham: Rowman & Littlefield Publishers, 2004.
Meister, Morris. *Children's Science Fair of the American Institute: A Project in Science Education*. New York: The American Institute and the American Museum of Natural History, 1932.
———. "Guiding and Aiding the Pupil in His Project." *General Science Quarterly* 3 (May 1919): 209–215.
———. *Living in a World of Science: Water and Air*. New York: Charles Scribner's Sons, 1930.
———. "Managing a Science Club." *School Science and Mathematics* 13 (March 1923): 205–217.
———. "Obituary: Otis William Caldwell, 1869–1947." *Science* 106 (December 12, 1947): 576–578.
———. "Outlook in Science Education." *The Science Teacher* 15 (October 1948): 105–107, 132.
———. "Pupil Adventure in Science." *High Points in the Work of the High Schools of New York City* 18 (September 1936): 5–12.
———. "Science in Elementary and Secondary Education." *The Science Teacher* 15 (February 1948): 13–16.
———. "Science Work in the Speyer School." *General Science Quarterly* 2 (May 1918): 429–445.

———. "The Educational Value of Certain After-School Materials and Activities in Science." PhD diss., Columbia University, 1921.

———. "The Junior Science Clubs: An American Institute Project in Science Education." *Science Education* 18 (April 1934): 68–74.

———. "We Have Lost a Generation." *The Science Teacher* 14 (April 1947): 62.

Midgette, Nancy Smith. *To Foster the Spirit of Professionalism: Southern Scientists and State Academies of Science*. Tuscaloosa: University of Alabama Press, 1991.

Miller, Donald W. "Suggested Programs for a Science Club." *Science Education* 14 (November 1929): 331–334.

Miller, Julie Ann. "Science Education for Whom?" *BioScience* 39 (June 1989): 354.

Mintz, Steven. *Huck's Raft: A History of American Childhood*. Cambridge, MA: Harvard University Press, 2004.

"Minutes of the Fifteenth Annual Meeting of the Alabama Junior Academy of Science, May 6, 1949." *Journal of the Alabama Academy of Science* 21 (February 1952): 71–72.

"Minutes of the Thirteenth Annual Meeting of the Alabama Junior Academy of Science." *Journal of the Alabama Academy of Science* 19 (December 1947): 90.

Moe, H. A. "Report of the Committee on Discovery and Development of Scientific Talent." In *Science—The Endless Frontier*, ed. Vannevar Bush, 136–185. Washington, DC: U.S. Government Printing Office, 1945.

Morgan, Banner Bill. *Transactions of the Wisconsin Academy of Sciences, Arts and Letters*. Madison, WI: Atwood & Culver, 1945.

Murphy, Frank W. "Science Clubs that Work." *General Science Quarterly* 4 (January 1920): 330–334.

National Academy of Sciences, National Academy of Engineering, and Institute of Medicine. *Rising above the Gathering Storm: Energizing and Employing America for a Brighter Economic Future*. Washington, DC: National Academies Press, 2007.

———. *Rising above the Gathering Storm Revisited: Rapidly Approaching Category 5*. Washington, DC: National Academies Press, 2010.

National Science Board. *America's Pressing Challenge—Building A Stronger Foundation. A Companion to Science and Engineering Indicators—2006*. Arlington, VA: National Science Foundation, January 2006.

National Science Teachers Association. *Science Instruction for National Security: 1945 Yearbook*. Washington, DC: National Science Teachers Association, 1945.

Noll, Victor H. *The Teaching of Science in Elementary and Secondary Schools*. New York: Longmans, Green, and Co., 1942.

NSTA. *Time for Science Instruction*. Washington, DC: NSTA, 1946.

Obama, Barack. "Remarks by the President at White House Science Fair." October 18, 2010. http://www.whitehouse.gov/the-press-office/2010/10/18/remarks-president-white-house-science-fair (accessed June 13, 2011).

Oerlein, Karl F. "Junior Academies of Science." *Educational Outlook* 14 (November 1939): 9–20.
———. "Science Clubs for Service." *School Science and Mathematics* 31 (March 1931): 314–320.
Oesterling, H. Carl. "The Illinois Junior Academy of Science." *School Science and Mathematics* 31 (April 1931): 461–463.
Owen, David. "Democracy." In *Political Concepts*, ed. Richard Bellamy and Andrew Mason, 105–117. Manchester and New York: Manchester University Press, 2003.
Palmeri, Joann. "An Astronomer beyond the Observatory: Harlow Shapley as Prophet of Science." PhD diss., University of Oklahoma, 2000.
Pape, Mary Elizabeth. "The Science Club." *School Science and Mathematics* 26 (May 1926): 552–554.
Patterson, James T. *Grand Expectations: The United States, 1945–1974*. New York: Oxford University Press, 1996.
Patterson, Margaret E. "Junior Scientists Assemble in Chicago." *The Science Teacher* 15 (February 1948): 27–29.
———. "Opportunity for Nation's Junior Scientists." *The Science Teacher* 17 (October 1950): 124–125.
———. "Science Club Members Profit from State and National Cooperation." *The Science Teacher* 14 (October 1947): 124–127.
———. "Science Clubs Expand Their Activities." *The Science Teacher* 15 (October 1949): 129–131.
Patterson, Margaret E., and Joseph H. Kraus. *Thousands of Science Projects*. Washington, DC: Science Service, 1956.
Pauly, Philip J. *Biologists and the Promise of American Life: From Meriwether Lewis to Alfred Kinsey*. Princeton: Princeton University Press, 2000.
———. "The Development of High School Biology: New York City, 1900–1925." *Isis* 82 (December 1991): 662–688.
Perrett, Geoffrey. *Days of Sadness, Years of Triumph: The American People, 1939–1945*. New York: Coward, McCann, & Geoghegan, 1973.
Perry, Carl F. "High School Science Fairs Are Worth the Effort." *School and Community* 42 (March 1956): 14–15.
Phares, Tom K. *Seeking—and Finding—Science Talent: A 50-year History of the Westinghouse Science Talent Search*. Pittsburgh: Westinghouse Electric Corporation, 1990.
Polenberg, Richard. *War and Society: The United States, 1941–1945*. Philadelphia: J. P. Lippincott, 1972.
Pollin, Burton R. *Toward Excellence in Education: Writings in Honor of Dr. Morris Meister*. Mission, KS: Inter-Collegiate Press, 1966.
Powers, Philip N. "The Changing Manpower Picture." *The Scientific Monthly* 70 (March 1950): 165–171.
Prewitt, Kenneth. "Scientific Illiteracy and Democratic Theory," *Daedalus* 112 (Spring 1983): 49–64.
Pricer, J. L. "Illinois State Academy of Science." *Science* 51 (March 26, 1920): 327.

"Program of the Winter Meeting." *Indiana Academy of Science Proceedings* 55 (1945): vii–xxiii.
Pruitt, Clarence M. "Activities of Chemistry Clubs." *Journal of Chemical Education* 4 (August 1927): 1037–1042.
Quarles, Donald A. "Cultivating Our Science Talent—Key to Long-term Security." *The Scientific Monthly* 80 (June 1955): 352–355.
Rader, Karen A., and Victoria E.M. Cain. "From Natural History to Science: Display and the Transformation of American Museums of Science and Nature." *Museum and Society* 6 (July 2008): 152–171.
Ransom, Sarah Bent. "The Science Fair as an Aid to Project Teaching." *Science Education* 22 (March 1938): 138.
Ravitch, Diane. *The Great School Wars: New York City, 1805–1973*. New York: Basic Books, 1974.
"Report of Committee on Science Talent Search Scholarships." *Transactions of the Illinois State Academy of Science* 40 (May 1947): 239.
"Report of the High School Relations Committee." *Iowa Academy of Science Proceedings* 62 (April 1955): 43.
"Report of the High School Relations Committee of 1953." *Iowa Academy of Science Proceedings* 60 (April 1953): 37.
"Report of the High School Relations Committee—1939." *Proceedings of the Iowa Academy of Science for 1939* 46 (April 1939): 38–40.
"Report of the High School Relations Committee 1939–1940." *Proceedings of the Iowa Academy of Science for 1940* 47 (April 1940): 27–29.
"Report of the Junior Academy Committee." *Proceedings of the West Virginia Academy of Science* 21 (December 1949): 13.
Rhees, David J. "A New Voice for Science: Science Service under Edwin E. Slosson, 1921–30." MA thesis, University of North Carolina, 1979.
Robinson, L. A. "The Physics Club in a Normal School." *School Science and Mathematics* 7 (June 1907): 461–462.
Rossiter, Margaret W. *Women Scientists in America: Before Affirmative Action, 1940–1972*. Baltimore: Johns Hopkins University Press, 1995.
Rudolph, John L. "Epistemology for the Masses: The Origins of 'The Scientific Method' in American Schools." *History of Education Quarterly* 43 (Fall 2005): 341–376.
———. "From World War to Woods Hole: The Use of Wartime Research Models for Curriculum Reform." *Teachers College Record* 104 (March 2002): 212–241.
———. "Portraying Epistemology: School Science in Historical Context." *Science Education* 87 (January 2003): 64–79.
———. *Scientists in the Classroom: The Cold War Reconstruction of American Science Education*. New York: Palgrave, 2002.
Rury, John L., and Shirley A. Hill. *The African American Struggle for Secondary Schooling, 1940–1980: Closing the Graduation Gap*. New York: Teachers College Record, 2012.
Rydell, Robert W. "The Fan Dance of Science: American World's Fairs in the Great Depression." *Isis* 76 (December 1985): 525–542.

Schrecker, Ellen. *Many Are the Crimes: McCarthyism in America*. Princeton: Princeton University Press, 1998.

Science Clubs of America. *How You Can Search for Science Talent: A Book of Facts about the Sixth Annual Science Talent Search for Westinghouse Scholarships*. Washington, DC: Science Service, 1946.

———. *Science and the Future: Essays of the Winners of the Westinghouse Science Scholarships in the Second Annual Science Talent Search*. Washington, DC: Science Clubs of America, 1943.

———. *Scientists of Tomorrow: Essays of the Winners of the Westinghouse Science Scholarships in the Fourth Annual Science Talent Search*. Washington, DC: Science Service, 1945.

———. *Scientists of Tomorrow: Essays of the Winners of the Westinghouse Science Scholarships in the Third Annual Science Talent Search*. Washington, DC: Science Service, 1944.

———. *Sponsor Handbook for 1958*. Washington, DC: Science Service, 1957.

"Science Exhibit of the American Institute." *School Science and Mathematics* 39 (March 1939): 225.

Science Service. *Youth Looks at Science and War: A Collection of Essays by the Washington Trip Winners of the First Annual Science Talent Search Conducted by Science Clubs of America*. Washington, DC: Science Service and Penguin Books, 1942.

Shapley, Harlow. "Status Quo or Pioneer? The Fate of American Science." *Harper's Magazine* 191 (September 1945): 312–317.

———. "The Scientist Outside the Laboratory." *The American Scholar* 15 (Autumn 1946): 411–415.

Sheldon, H. H. "The Science Club Program of the American Institute." *School Science and Mathematics* 40 (April 1940): 365–367.

Shreve, Robert P. "Warning—Proceed With Caution." *The Science Teacher* 24 (November 1957): 334.

Slaton, Amy E. *Race, Rigor, and Selectivity in U.S. Engineering: The History of an Occupational Color Line*. Cambridge, MA: Harvard University Press, 2010.

Smith, Guy M. "Science Clubs in the High School." *School Science and Mathematics* 25 (October 1925): 720–725.

Snyder, Emily Eveleth. "Report on the Biology Club of the Little Falls High School." *School Science and Mathematics* 31 (January 1931): 32–33.

Spring, Joel. *The Sorting Machine: National Educational Policy Since 1945*. New York: David McKay Company, 1976.

Stanford University School of Education Faculty. *Education in Wartime and After*. New York: D. Appleton-Century Company, 1943.

Studebaker, John W. "What the Secondary Schools Can Do to Help Win This War." *Bulletin of the National Association of Secondary-School Principals* 26 (October 1942): 11–17.

Sugrue, Thomas J. "Reassessing the History of Postwar America." *Prospects* 20 (1995): 493–509.

Susman, Warren I. "The People's Fair: Cultural Contradictions of a Consumer Society." In *Dawn of a New Day: The New York World's Fair, 1939/40*, ed. Helen A. Harrison, 16–27. New York: New York University Press, 1980.
Symington, W. Stuart. "Aids National Security." *Science News Letter* 59 (April 7, 1951): 221.
"Talent Searches—State and National." *Journal of the Tennessee Academy of Science* 25 (October 1950): 298–307.
Taylor, Mary. "Teaching Science through the Children's Fair." *American Childhood* 15 (March 1930): 6–9, 58–59.
Terzian, Sevan G. "'Adventures in Science': Casting Scientifically Talented Youth as National Resources on American Radio, 1942–1958." *Paedagogica Historica* 44 (June 2008): 309–325.
———. "*Science World*, High School Girls, and the Prospect of Scientific Careers, 1957–1963," *History of Education Quarterly* 46 (Spring 2006): 73–99.
"The American Institute and the Junior Science Clubs." *School and Society* 41 (February 9, 1935): 185–186.
The American Institute of the City of New York. *How to Organize a Science Club*. New York: The American Institute, 1938.
"The Children's Science Fair of the American Institute, New York." *School and Society* 41 (April 13, 1935): 502.
The Thirty-first Yearbook of the National Society for the Study of Education: Part I, a Program for Teaching Science. Bloomington, IL: Public School Publishing Company, 1932.
Thomson Jr., John W. "How the Junior Academies of Science Operate." *The Scientific Monthly* 64 (April 1947): 327–336.
Tobey, Ronald C. *The Ideology of National Science, 1919–1930*. Pittsburgh: University of Pittsburgh Press, 1971.
Tolley, Kim. *The Science Education of American Girls: A Historical Perspective*. New York: RoutledgeFalmer, 2003.
Tyack, David B. *The One Best System: A History of American Urban Education*. Cambridge, MA: Harvard University Press, 1974.
Tyack, David B., Robert Lowe, and Elisabeth Hansot, *Public Schools in Hard Times: The Great Depression and Recent Years*. Cambridge, MA: Harvard University Press, 1984.
Urban, Wayne J. *More Than Science and Sputnik: The National Defense Education Act of 1958*. Tuscaloosa: University of Alabama Press, 2010.
Walton, A.C. "The Des Moines Session of the Academy Conference." *Science* 71 (February 7, 1930): 147–148.
Wang, Jessica. *American Science in an Age of Anxiety: Scientists, Anticommunism, and the Cold War*. Chapel Hill: University of North Carolina Press, 1999.
———. "Liberals, the Progressive Left, and the Political Economy of Postwar American Science: The National Science Foundation Debate Revisited." *Historical Studies in the Physical and Biological Sciences* 26, no. 1 (1995): 139–166.

———. "Scientists and the Problem of the Public in Cold War America, 1945–1960." *Osiris* 17 (2002): 323–347.
Ward, W. H. "South Carolina Junior Academy of Science." *Bulletin of the South Carolina Academy of Science* 15 (1953): 4.
Waterman, Alan T. "Role of the Federal Government in Science Education." *The Scientific Monthly* 82 (June 1956): 286–293.
Weller, Florence, *et al.* "A Survey of the Present Status of Elementary Science." *Science Education* 17 (October 1933): 193–198.
Wendt, Gerald. *Science for the World of Tomorrow.* New York: W. W. Norton & Company, 1939.
Wessel, Thomas, and Marilyn Wessel. *4-H: An American Idea, 1900–1980.* Chevy Chase, MD: National 4-H Council, 1982.
Westbrook, Robert. *John Dewey and American Democracy.* Ithaca: Cornell University Press, 1991.
———. "Public Schooling and American Democracy." In *Democracy, Education, and the Schools*, ed. Roger Soder, 125–150. San Francisco: Jossey-Bass Publishers, 1996.
Wile, Frederic William. *A Century of Industrial Progress.* New York: Doubleday, Doran & Company, 1928.
Williams, M. M. "The Junior Academy of Science, Its Present Organization and Future Possibilities of Our State and Nation." *Teachers College Journal* 5 (September 1933): 148–152.
Wright, Stephen J. "Impact on the High-school Curriculum." *Journal of Educational Sociology* 16 (March 1943): 424–450.
Wyatt, W. W. "The Tennessee Junior Academy of Science." *Journal of the Tennessee Academy of Science* 30 (April 1955): 141–150.
Xan, John. "Alabama State Science Talent Search." *Journal of the Alabama Academy of Science* 20 (December 1948): 96.
Yarbrough, John A. "North Carolina Science Fairs and the North Carolina Academy of Science." *The High School Journal* 39 (February 1956): 270–274.
Zilversmit, Arthur. *Changing Schools: Progressive Education Theory and Practice, 1930–1960.* Chicago: University of Chicago Press, 1993.
Zim, Herbert. "Junior Scientists Assembly." *The Science Teacher* 14 (February 1947): 24–27, 42.
Zim, Larry. *The World of Tomorrow: The 1939 New York World's Fair.* New York: Harper & Row, 1988.

INDEX

Academies of Science:
Alabama, 123, 198 (n. 92); clubs & memberships, 167 (n. 58); college scholarships, 120–4; Georgia, 121; Illinois, 17–19, 121, 154 (n. 51); Indiana, 19, 52, 53–4, 121, 124, 127, 154 (n. 51); Iowa, 19, 52, 75, 122, 124; Kentucky, 153 (n. 37); Louisiana, 122; Michigan, 88; Nebraska, 53; North Carolina, 75–6; Oklahoma, 55, 169 (n. 72); Pennsylvania, 53, 75, 124; South Carolina, 124; South Dakota, 121; Tennessee, 122, 127, 199 (n. 97); Virginia, 121, 124; Washington (DC), 123–4
Academy Conference, 52–3, 121
Adventures in Science: national shortage of scientists, 115; science for a peaceful world, 95, 105, 116; students for wartime service, 86, 87, 91, 94; student speakers, 92, **117**
Alabama Academy of Science. *See* Academies of Science
Albion College, 88
America's Pressing Challenge—Building a Stronger Foundation, 143
American Association for the Advancement of Science (AAAS), 5, 16, 34, 61, 119, 142
American Cancer Society, 87, 110
American Chemical Society, 43

American Competitiveness Initiative: Leading the World in Innovation, 143–4
American Institute of the City of New York:
beyond New York City, 52–3, 54–6, 62–3; Christmas Lectures, 40, 123; Crime Prevention Bureau, 43–4; demonstration lectures, 37; early history, 21–2, 155 (n. 49); museum and workshop courses, 6, 35, 36–8, 41, 51, 59; New York World's Fair, 6, 57–61, 63–5, 66–75; science clubs, 33–44, 51–2; science congresses, 38–9, **41**; science fairs, 5, 23, 24–6, 29–31, 44–50, 125, 130; and Science Service, 82–3, 85, 87; and Westinghouse, 6, 60–2, 69–73, 75–8
American Institute Science & Engineering Clubs. *See* Junior Science Clubs
American Museum of Natural History:
Christmas Lectures, 40; demonstration lectures, 37; museum and workshop courses, 6, 36–8; science fairs, 22, 24, 34, 44, 49, 54, 71, 125, 139
American Science Teachers Association, 84

Astell, Louis, 17–20, 30, 154 (n. 41)
atomic age:
   citizenship in, 109, 144; Cold War, 3, 101, 139; expert leadership, 7, 120, 138; science education in, 103, 104; US political power, 1
atomic technology, 106, 113, 115–16, 127
atomic weapons, 103–4, 112, 114, 115, 118
Axelrod, Julius, 143

Baker, Ross A., 36
Barker, J. W., 93
Barker, James P., 37
Bilsing, S. W., 53
Blakeslee, Albert, 58
Blanchard, Kenneth, 37
Board of Education of the City of New York, 35, 50, 60, 73, 125
Bodet, Jaime Torres, 111
Boehmer, Kathryn M., 123
Branch, Hazel, 18–19
Britt, Steuart Henderson, 90–1, 99, 131–3
Bronx High School of Science (BHSS), 64, 92, 106, 152 (n. 17), 191 (n. 25)
Brooklyn Children's Museum, 23, 37, 38
Buffalo Museum of Science, 29
Buhl Planetarium and Institute of Popular Science, 125, 126
Burnham, John C., 53, 72
Bush, Vannevar, 96, 100, 105

Caldwell, Otis:
   American Institute, 25; Junior Academies, 19, 52, 53; science fairs, 27–8; on scientific progress & citizenship, 20, 30
Camp Fire Girls, 83
Campbell, Harold G., 64
Cannon, J. K., 85

Cardinal Principles of Secondary Education, 10
Carmichael, Leonard, 93
Carnegie Institute of Technology, 83, 89
Carnegie Institution, 36, 38, 58, 118
Carson, John, 141
Caryl B. Haskins, 118
CBS radio, 51, 86.
   See also *Adventures in Science*
Chambers, Robert, 40
Children's Science Fair.
   See science fairs
Christmas Lectures, 35, 40, 50, 55, 59, 123
Churchill, Winston, 116
City College (of New York), 36, 150 (n. 3)
Clark, James P., 37
Clark, John A., 36
Clinton Corn Processing Company, 124
Coit, Marjorie, 23–4, 27, 28–9
Cold War:
   atomic weapons, 103–4, 112, 114, 115, 118; political ideology, 4, 7, 105, 112, 119; science talent, 78–9, 114, 120, 130, 139, 140, 143, 144; security restrictions, 116–17.
   See also Sputnik
College Entrance Examination Board, 140
college scholarships:Academies of Science, 120–4; James Bryant Conant, 97–8; Watson Davis, 90; for national service, 94, 112; proposed federal plans, 98–100; science fairs, 125, 128; Science Talent Search, 83, 88–9, 93; and unequal opportunities, 140–1; Westinghouse, 89, 114
Columbia University, 9, 28, 36, 93

Commission for the Reorganization of Secondary Education, 11
Committee on the Place of Science in Education, 19, 52
Compton, Arthur, 127
Compton, Karl T., 94
Conant, James Bryant, 96–8, 137
Condon, E. U., 63, 94–5, 115–16
Conn, Steven, 36, 65
Cornell University, 24, 93
Cottam, Clarence, 115
Craig, A. P., 65
Creager, Mary, 121
Crime Prevention Bureau, 43
Crown Cork and Seal Company, 87
Cunningham, Bert, 76
Cusker, Joseph, 78

Davis, Watson:
  Academies of Science, 121–2, 123; on atomic technology, 103, 104; Cold War, 115, 118, 130–1, 140; National Science Fair, 126, 129–30; New York World's Fair, 58; professional networks, 82–3, 179 (n. 85), 180 (n. 3), 184 (n. 140), 188 (n. 98), 193 (n. 51); on science education for democracy, 95–6, 107–8, 110–11, 129, 130; on science education in World War II, 84–5, 86–8, 90–3, 94; Science Talent Search, 6, 90–3, 94, 100, 106, 115, 118, **120**, 121–2, 132–3, 190 (n. 19), 203 (n. 132); *Science—The Endless Frontier*, 97, 98, 100, 188 (n. 98); in support of a national science foundation, 106–8, 190 (n. 19); UNESCO, 110–11, 192 (n. 42). See also *Adventures in Science;* Science Service
democratic citizenship:
  John Dewey's conceptions of, 3–4, 144, 151 (n. 14); educational ideals, 2–3, 101, 141, 144–5;
Morris Meister's conceptions of, 13, 42, 65, 108–9; New York World's Fair, 59, 61, 65, 74, 76, 78–9; J. Robert Oppenheimer, 117; relations to meritocracy, 104, 107–8, 111, 118–19, 130, 141; and science education, 5, 44, 52, 53, 84, 103, 139, 140; Harlow Shapley, 116; Alan Waterman, 129; Watson Davis, 95–6, 107–8, 110–11, 129, 130
demonstration lectures, 34, 36–7, 161 (n. 13)
Dewey, John:
  on democracy, 3–4, 144; William Heard Kilpatrick, 12, 151 (n. 15); Morris Meister, 5, 9, 12–13, 15, 151 (n. 15); pedagogy, 12, 18, 28, 59; as a pragmatist, 4
Ditmars, Raymond L., 36
Domestic Policy Council, 143–4
Dorn, Charles, 101
Dow Chemical Company, 124
draft deferments, 93, 105, 107, 190 (n. 16)
DuPont, 58

Edgerton, Harold A., 90–1, 99, 131–4
Eisenhower, Dwight D., 139, 140
Eli Lilly Company, 129
Elizabeth Peabody House, 54
Emig, Catherine, 42, 48–9
Enders, Howard E., 53
Ernst, Frederic, 71–2
Evander Childs High School, 24, 34

Farr, Wanda K., 37
Fass, Paula, 48
Federation of Science Teachers Association of New York, 125
Federer Jr., Charles, 62

female students:
  American Institute student laboratory, 77; Children's Science Fair, 45, **47–9**; 4-H Clubs, 15; on marriage, 133–5; National Science Fair, 129; New York World's Fair, 66–7; Science Congress, 38, 39, 63; in science courses, 48, 135; Science Talent Search, 92, 132–5, 138, 140, 141, 204 (n. 136)
Force, Edith, 55
4-H Clubs, 15, 16, 83, 89
Franklin Institute, 126
Furer, J. A., 94
Future Farmers of America, 113

General Electric Company, 38, 58, 114, 171 (n. 19)
General Motors, 67, 125
General Science Association of New York, 90
Georgia Junior Academy of Science. *See* Academies of Science
Girls Commercial High School, 38, 66
Great Depression:
  American Institute, 25, 29, 33; impact on schools, 35, 50; New York World's Fair, 66; societal problems, 5, 52; and women, 49; World War II, 78

Hartman, Andrew, 130
Harvard University, 40, 83, 96, 98, 104, 110, 133
high school enrollments. *See* student enrollments
High School Science Section (Illinois). *See* Academies of Science
High School Section of the Nebraska Academy of Science. *See* Academies of Science

Hunter, George W., 17
Hutcheson, J. A., 115
Hutchins, L. W.:
  on civic benefits of science education, 40–1, 43, 82; conflict within American Institute, 25, 50; educational programs, 22, 26, 37, 40, 49; as popularizer of science, 21–2

Illinois State Academy of Science. *See* Academies of Science
Indiana Academy of Science. *See* Academies of Science
Intel Corporation, 142, 208 (n. 12)
International Business Machines Corporation (IBM), 76
Iowa Junior Academy of Science. *See* Academies of Science

Jet Propulsion Laboratories, 130
Jewett, Andrew, 144
John Simon Guggenheim Memorial Foundation, 96
Johnson, George E., 18
Johnson, Leroy, 116
Johnson, Philip G., 108
Jones, Norman R. D., 125
Junior Chamber of Commerce of the United States, 70
Junior Red Cross, 81
Junior Science Clubs:
  new programs, 35–44, 51–2; New York World's Fair, 57–8, 63–5, 66–70, 73–4; origins, 33–5
Junior Science Hall. *See* New York World's Fair
Junior Scientists Assembly, 120–1
juvenile delinquency, 2, 10, 14, 18, 20, 33, 43

Kassner, James L., 123, 198 (n. 92)
Kelly, Agnes G., 49
Kilgore, Harley, 106, **107**

Kilpatrick, Van Evrie, 23, 27
Kilpatrick, William Heard, 12, 151 (n. 15)
Knight, Alfred, 34–5
Knox, William Franklin, 85
Knox College, 17
Kohlstedt, Sally Gregory, 22
Korean War, 104, 118, 129
Kraus, Joseph, 85, 128
Kuznick, Peter J., 51, 66

Lark-Horovitz, Karl, 115
Leech, Edward T., 88
Louisiana Academy of Science. See Academies of Science
Lufkin, Hoyt D., 58, 60, 63, 68, 72

MacCallum, Hazel, 71
Mann, Albert Russell, 24
Mann, Paul B., 24, 34, 36, 61, 64
Marchand, Roland, 65
Massachusetts Institute of Technology, 94, 113
McEvoy, S. Aleta, 17–18
McTyeire, Clustie, 123
Mead, Margaret, 95
Meister, Morris:
    Academies of Science, 55, 120–1; Bronx High School of Science, 64, 152 (n. 17); Children's Science Fair, 26, 28, 29–31, 50, 125; Crime Prevention Bureau, 43; as critic of nature study, 159 (n. 88); early influences, 150 (n. 3); Junior Science Clubs, 33–6, 39–40, 41–2, 61, 62; on juvenile delinquency, 9–12, 20; National Science Teachers Association, 108–9; as a pioneer of science clubs, 5, 9–15, 16; science education for democracy, 5, 64, 95, 108–9, 110, 139, 140. See also American Institute of the City of New York; democratic citizenship;
John Dewey; National Science Foundation
meritocracy:
    Cold War, 104, 119, 139; James Bryant Conant, 96–8; ideals in American education, 2–3, 4–5, 141, 148 (n. 10); and quest for scientific talent, 6–7, 79, 90, 104, 107–8, 109, 111, 115; science fairs, 130–1, 139; Science Talent Search selection methods, 131–8; *Science—The Endless Frontier*, 96–100; World War Two, 6–7, 79, 90
Meyerhoff, Howard, 130. See also Scientific Manpower Commission
Michigan Junior Academy of Science. See Academies of Science
military deferments. See draft deferments
Miller, E. C. L., 120
Millet, Ralph T., 85
Mintz, Steven, 135
Mitchell, James P., 87–8
Moe, Henry Allen, 96, 97–8, 99–100
Monteith, A.C., **120**
Morrison, A. Cressy, 25
Mulrooney, Edward P., 43
Murdock, Edwin F., 24–5
museum and workshop courses, 6, 35, 37–8, 40, 41, 51, 59
Museum of Science and Industry (Chicago), 54
Museum of Science and Industry (New York), 36, 37

Nardroff, Ernest von, 38
National Academies (of Science), 143
National Academy of Science, 16, 20, 52

National Defense Education Act, 137, 141
National Inventors Council, 92–3
National Research Council's Office of Scientific Personnel, 108
National Science Board (NSB), 143, 208 (n. 14)
National Science Fair:
  industrial support of, 128–9, 142, 201 (n. 112); military sponsorship, 129–30; for national strength, 3, 7, 104, 126–31, 139; newspaper sponsorship, 125–**8**.
  See also science fairs
National Science Foundation:
  Alan Waterman, 129, 140; Morris Meister's support of, 107; National Science Fair, 129–30; Science Service, 112; Science Talent Search, 106, **107**, 132; *Science—The Endless Frontier*, 96, 100, 105; State Academies of Science, 124; Watson Davis's support of, 106–8, 190 (n. 19)
National Science Teachers Association (NSTA), 108, 180 (n. 3), 188 (n. 98)
nature study:
  Children's Fairs, 23, 25–6, 28; criticisms of and decline, 26–7, 29–30, 48; popularity, 20, 22–3, 43; science clubs, 16, 20.
  See also science curricula
Naval Research Laboratory, 93
Nebraska State Teachers Association, 53
New Jersey Department of Agriculture, 54–5
New York Academy of Public Education, 96
New York Association of Biology Teachers, 51
New York Principals Association, 51

New York University, 6, 21, 37, 51, 64
New York World's Fair:
  American Institute, 58–61, 64–5, 68–9, 71, 75–7; conflicts in the presentation of science, 6, 66–7, 71–3; Junior Science Hall, 57, 65, 66–7, **69**, 71; student participation in, 57, 63, **69**, 73–7; Student Science Labs, 69, 74; Westinghouse, 6, 57–8, 60–3, 65–7, 69–70, 71–3, 75–6, 142, 144, 170 (n. 2); World War II, 74–5
Newcomb, H. T., 34–5
Newman, Hugo, 1, 29, 150 (n. 3)
Newtown Agricultural High School, 24, 25
North Carolina Academy of Science. *See* Academies of Science
Northrup, Alice Rich, 23

O'Connor, Basil, 116
Oak Ridge Institute of Nuclear Studies, 106, 124, 127, 133, 201 (n. 112)
Oerlein, Karl F., 53
Office of Defense Health & Welfare Services, 94
Ohio State University, 90
Oklahoma Academy of Science. *See* Academies of Science
Oppenheimer, J. Robert, 116–17, **120**
Owen, Russell, 40

Parmelee, H. C., 64, 73–4, 76
Parr, Rosalie M., 17
Parran, Thomas, 95
Patterson, Margaret, 83, 105, 111, 112, 121, 127, 128
Pauly, Philip, 17, 22
Pendray, G. Edward, 40, 82–3

Pennsylvania Junior Academy of Science. *See* Academies of Science
Physics Teachers Club of New York, 51
Piccard, Jean, 40
Pollock, Robert T., 61, 63, 68–9, 172 (n. 26)
Prescott, G. W., 88
Prewitt, Kenneth, 144
Price, Gwilym A., 114, 115
Pricer, J. L., 17
project method, 5, 11, 12, 13, 14, 17, 30
Purdue University, 53

Quarles, Donald A., 118

Radcliffe College, 93, 133
Rhode Island School of Design, 54
Riddle, Oscar, 36, 156 (n. 54)
Rising Above the Gathering Storm, 143
Ritter, William E., 78
Roosevelt, Franklin D., 96, 98
Rossiter, Margaret, 134
Royal Institution of London, 40
Rudolph, John L., 4, 12, 110
Rydell, Robert W., 66–7

School Nature League, 22, 23, 25, 27, 29
Science Clubs of America (SCA):
 Watson Davis, 109–10; growth of, 83, 109, 112–13; search for scientific talent, 3, 100, 105; Harlow Shapley, 97; in World War II, 85–8, 93
science clubs—origins:
 early examples, 15–16; and Junior Academies of Science, 17–21; Morris Meister, 9–15
science congresses, 6, 35, 38–41, **41**, 42, 45, 50, 53, 54, 62, 63, 64

science curricula:
 biology, 16, 17–18, 19, 22, 23, 29–30, 84, 135, 136; chemistry, 16, 19, 27, 39, 48, 84, 135, 136, 137; nature study, 16, 20, 22–4, 25, 26–7, 28, 29–30, 43, 48; physics, 16, 19, 20, 27, 39, 48, 84, 135, 136–7
science fairs:
 Children's Science Fair, 5, 21–30, 33, 33–5, 44–5, **46–7**, 47–52, 55, 56, 71, 77, 164 (n. 31), 165 (n. 33), 200 (n. 100); girls' participation in, 45, 47–8; National Science Fair, 3, 7, 104, 126–**8**, 128–9, 129–31, 139, 142, 201 (n. 112); in various municipalities, 54–6, 88, 112, 123–4, 125–7, 159 (n. 87), 168 (n. 64), 199 (n. 93);White House, 208 (n. 11). *See also* New York World's Fair
Science for All Americans, 142
Science Leaflet, 43, 70
Science News Letter, 85, 87, 88, 89, 90, 93, 100, 112
Science Observer, 57, 62, 68, 74, 75, 76
Science Service:
as agency for popularizing science, 78, 82–3, 95; on atomic technology, 103–5; National Science Fair, 126–8; New York World's Fair, 58; science clubs, 112–13; *Science—The Endless Frontier*, 97–101, 105; Society for Science and the Public, 150 (n. 18); State Academies of Science, 121–2, 123, 124; UNESCO, 110–11, 192 (n. 42); Westinghouse, 77, 83, 89–90, 114; World War II, 81, 84–8. *See also* Science Clubs of America (SCA); Science Talent Institute; Science Talent Search

Science Talent Institute, 91, 93, 94, 107, 113, 115, 116, **117, 119**
Science Talent Search:
  Cold War, 7, 104, 112, 116–19; Congressional recognition of, 106–7; follow-up studies, 132–5; gender differences, 132–5, 138, 140, 141, 204 (n. 136); Intel Corporation, 208 (n. 12); methods of selection, 90–1, 131–2, 137–8, 141; as a national resource, 105, 107, 111, **117, 119, 120**, 138–9, 142; origins, 2, 6, 78, 82–3; as precedent, 96–100; regional differences, 135–7; state cooperation, 119–24; Westinghouse, 6–7, 78, 88–90, 113–14, 115, **120**, 142; World War II, 3, 88–90, 91–5, 101, 143–4. *See also* Science Service; scientific talent; Science Talent Institute
Science—The Endless Frontier, 7, 96–100, 105
Scientific Manpower Commission, 130
scientific talent:
  Congressional considerations of, 106–7; for democracy, 61; for international peace, 114; for military strength, 87, 92, 114–15, 117–19, 129–30; as a national resource, 87, 96–100; projected deficits, 96–100, 114–15, 120, 127, 141. *See also* Science Talent Institute Science Talent Search
Scripps, E. W., 78
Seaborg, Glenn, 127
Sears, Paul B., 115
Shapley, Harlow:
  Christmas Lectures, 40; on draft deferments, 190 (n. 16); on science as international, 104–5, 110, 116, 190 (n. 13), 192 (n. 39), 197

(n. 82); Science Talent Search, 83, 115, **120**, 140; *Science—The Endless Frontier*, 97, 98
Shaw, Robert P., 36
Sheldon, H. H., 21–2, 36, 64, 71–3, 74, 75, 82
Sinclair Research Laboratories, 134
Smith, Michael L., 65
Smith-Lever Act, 15
Smithsonian Institution, 16, 127
Smyth, Henry DeWolf,118
  See also US Atomic Energy Commission
South Carolina Academy of Science. *See* Academies of Science
South Dakota Junior Academy of Science. *See* Academies of Science
Soviet Union, 3, 7, 101, 103, 104, 105, 112, 114, 116, 130, 137, 139–40
Speyer Junior High School, 9, 12, 13–14
Sputnik, 2, 3, 130, 137, 139, 140, 143
Stanley, Wendell M., 115
State High School Chemistry Teachers' Association (Illinois), 17
student enrollments:
  girls in science courses, 48, 135; high school, 2, 3, 18, 33, 50, 81; in science curricula, 27, 48, 84, 153 (n. 36)
Student Science Labs. *See* New York World's Fair
Stuyvesant High School, 38, 68, 86, 136
Symington, W. Stuart, 1, 129

Teachers College, 1, 19, 51
Tennessee Academy of Science. *See* Academies of Science
Terman, Lewis, 132
Thomas, Lyell J., 17–18, 95, 122
Thone, Frank, 103, 105, **117**
Tolley, Kim, 39, 48, 49

Truman, Harry S., 98, 116, 118, **119**
Trytten, M. H., 108, 117
Tufts College, 93
Tuskegee Institute, 123

United Nations, 104
United Nations Educational, Scientific, and Cultural Organization (UNESCO), 110–11, 192 (n. 39 & 42)
University of Alabama, 123
University of Chicago, 113, 134
University of Illinois, 17, 18, 122
University of Iowa, 129
University of Oklahoma, 121
University of Wichita, 18
Urey, Harold C., 40
US Atomic Energy Commission, 104, 112, 113, 118, 194 (n. 53)
US Department of Agriculture, 15, 37, 87
US Department of Defense, 112, 118, 129
US Fish and Wildlife Service, 87, 115
US Forest Service, 87
US Office of Education, 137
US Office of Scientific Research and Development, 7

Victory Corps, 81
Virginia Academy of Science. *See* Academies of Science
Virginia Education Association, 111
Vogt, William, 115

Wang, Jessica, 112
Washington (DC) Junior Academy of Science. *See* Academies of Science

Washington University, 127
Waterman, Alan T., 129, 140
Wendt, Gerald, 42, 51, 54, 58, 61, 164 (n. 26)
Westinghouse Educational Foundation, 89, 194 (n. 53 & 55)
Westinghouse Electric and Manufacturing Company: American Institute, 6, 60–2, 63, 69, 70–3, 75–7, 171 (n. 19); E.U. Condon, 94–5; New York World's Fair, 6, 57–8, 60–3, 65–7, 69–70, 71–3, 75–6, 142, 144, 170 (n. 2); other educational initiatives, 113–14, 194–5 (n. 56 & n. 57); Science Congress, 38; Science Service, 6, 77–8, 82–3, 89–90, 195 (n. 58); Science Talent Search, 6–7, 78, 88–90, 113–14, 115, **120**, 142; World War II, 88–90
Whitman, Walter G., 118
Wilson, M. L., 94
Wollaston Mothers' Club, 54
Work, Lincoln T., 36–7
World War I, 2, 9, 10, 15
World War II: impact on schools, 5, 81, 139, 140, 180 (n. 2); mobilization of science education, 3–4, 6, 78–9; political consequences of, 7, 103–5, 144; science clubs, 84–7; scientific talent, 87–96, 101, 107, 112, 117, 120, 143

Youngholm, David, 65

GPSR Compliance
The European Union's (EU) General Product Safety Regulation (GPSR) is a set of rules that requires consumer products to be safe and our obligations to ensure this.

If you have any concerns about our products, you can contact us on

ProductSafety@springernature.com

In case Publisher is established outside the EU, the EU authorized representative is:

Springer Nature Customer Service Center GmbH
Europaplatz 3
69115 Heidelberg, Germany

www.ingramcontent.com/pod-product-compliance
Lightning Source LLC
LaVergne TN
LVHW011007250326
834688LV00004B/115